NO ORDINARY MIKE

NO *Ordinary* MIKE

Michael Smith, Nobel Laureate

Eric Damer & Caroline Astell

Foreword by Richard J. Roberts

RONSDALE PRESS

NO ORDINARY MIKE: Michael Smith, Nobel Laureate

RONSDALE PRESS
3350 West 21st Avenue
Vancouver, B.C., Canada
V6S 1G7

Edited by Ronald B. Hatch
Typesetting: Julie Cochrane, in New Baskerville 11 pt on 15
Cover Design: Julie Cochrane
Cover Photo: Dina Goldstein

Ronsdale Press wishes to thank the Canada Council for the Arts, the Government of Canada through the Book Publishing Industry Development Program (BPIDP), and the Province of British Columbia through the British Columbia Arts Council for their support of its publishing program.

National Library of Canada Cataloguing in Publication

Damer, Eric, 1964–
No ordinary Mike: Michael Smith, Nobel Laureate / Eric Damer & Caroline Astell.

Includes bibliographical references and index.
ISBN 1-55380-014-1

1. Smith, Michael, 1932–2000. 2. Biotechnologists — Canada — Biography. 3. University of British Columbia — Faculty — Biography. 4. Biochemists — Canada — Biography. I. Astell, Caroline R., date II. Title.

QP511.8.S55D34 2004 660.6'092 C2004-901700-4

At Ronsdale Press we are committed to protecting the environment. To this end we are working with Markets Initiative (www.oldgrowthfree.com) and printers to phase out our use of paper produced from ancient forests. This book is one step towards that goal.

Printed in Canada by AGMV Marquis

ACKNOWLEDGEMENTS

As with any work such as this, many people provided useful help. In the background were those who provided salary support and funds for publishing. Donations were gratefully received from the Biotechnology Laboratory at UBC; the UBC Offices of Vice-President Research, and Vice-President Academic and Provost; the Michael Smith Foundation for Health Research; the Department of Biochemistry and Molecular Biology, UBC; ZymoGenetics Inc.; Dr. and Mrs. Earl Davie; and the Peter Wall Institute for Advanced Studies. Proceeds from the sales of this book will be used to fund graduate scholarships in Biochemistry, Molecular Biology and Biotechnology.

Many people shared their memories, interpretations, and opinions, including members of Michael Smith's family, his partner Elizabeth Raines, friends, co-workers, students, and scientific colleagues. Many thanks to contacts in England: Robin Smith, Dennis Smith, Joan Bielby, Peter Leeming, Colin Booth, Ian Wrigley, Roy

Whitehead, Harry Thompson, Sheila Groom, Fred Sanger, Nigel Brown, and Duncan McCallum. Those on this side of the Atlantic who contributed their time and memories include Gordon Tener, Bill Hoar, Jack Campbell, Edward M. Donaldson, Ulf Fagerlund, Ian Gillam, Gordon (Bob) Cherry, Peter Schmidt, Nadine Wilson, Shirley Gillam, Helen Smith, Philip Bragg, Victor Ling, Bill Cullen, David Frost, Richard Barton, Diana (Crookall) Bragg, Darlene Crowe, Phil Heiter, Jane Roskams, Bob Miller, Grant Mauk, Patricia and Walter Jahnke, Pat and Jack Wood, Vivian MacKay, W. Ford Doolittle, Judith Hall, Patricia Baird, Michael Harcourt, Jeanette Beatty, Heather Merilees, Gary Pielak, Dennis Luck, Barry McBride, John Ngsee, Elsie Wollaston, Gary McKnight, Elizabeth Raines, Tomoo Nakano, Sarb Ner, Hitoshi Ohtaki, Brian James, John Moffatt, Rosie Redfield, Jack and Margie McLellan, John Hobbs, Don Brooks, Roger Foxall, Marco Marra, Al and Irene Whitney, and Julian Davies. Unless otherwise documented, stories and anecdotes in the text come from interviews and personal correspondence with these people.

A number of people provided invaluable assistance with archives and other documents. In the Blackpool area, assistance was provided by Elaine Smith and Barry Shaw of the Blackpool Civic Trust, Simon Kularatne, Ted Lightbown, Alan Eaves, Christine Storey, and Kenneth Shenton. Andrew Thynne at the Lancashire Record Office located and transcribed archival records, while James Peters and Dave Smith found useful information in the University of Manchester Archives. Closer to home, Krista Kaptein helped locate a news story at the Comox Archives and Museum and Elizabeth Raines granted access to her private collection. Thanks also to the staff of the University of British Columbia Archives, particularly Lesley Field for his help with photographs and Chris Hives who provided administrative support. Every reasonable effort has been made to identify the owners of photographs and obtain permission for reproduction. Any further information will be gratefully received and acknowledged.

Historians William Bruneau and Jean Barman supplied useful comments on earlier drafts while biochemists Gordon Tener, Phil

Bragg, and John Hobbs checked the science. Ronald Hatch of Ronsdale Press provided his usual keen editing and good humour during the final stages. Apologies to anyone we have missed and thanks to all those who helped with important details but who are too numerous to list.

Despite the assistance we received, the authors bear the final responsibility for the choice and interpretation of evidence and the organization of this biography.

ABBREVIATIONS

BRC	Biomedical Research Centre
CFI	Canada Foundation for Innovation
CIAR	Canadian Institute for Advanced Research
CIHR	Canadian Institutes of Health Research
DNA	Deoxyribonucleic Acid
FRB	Fisheries Research Board of Canada
GMO	Genetically Modified Organism
MRC	Medical Research Council of Canada
NMR	Nuclear Magnetic Resonance
NRC	National Research Council
NSERC	Natural Sciences and Engineering Research Council
PENCE	Protein Engineering Network Centre of Excellence
RNA	Ribonucleic Acid
SCWIST	The Society for Canadian Women in Science and Technology

LIST OF SOURCES

(All Fonds and Collections in UBC Archives unless otherwise noted)

ABBREVIATION	COLLECTION
(none)	Smith Fonds
(none)	Faculty of Medicine Fonds
BL Files	Files kept at the UBC Biotechnology Laboratory (not accessioned)
Senate	UBC Senate Minutes
Scrapbook File	Smith Scrapbook File kept at the UBC Department of Biochemistry
Personnel File	Smith Personnel File kept at the UBC Department of Biochemistry
BoG	UBC Board of Governors
Raines Collection	Private Collection of Elizabeth Raines, Vancouver
UMDCA	University of Manchester, Department of Chemistry Archives
UMA	University of Manchester Archives
(none)	Lancashire Record Office

CONTENTS

FOREWORD

Dr. Michael Smith was an uncommon scientist who used synthetic organic chemistry to create an extraordinarily powerful tool that is now used by biologists worldwide. We often forget the remarkable influence that tools have on our ability to conduct research since so much more attention gets focused on the results of that research and the exciting findings that constantly seem to emerge from our studies of biology. Wisely, the Nobel Committees in Stockholm have frequently recognized the immense contributions of the tool developers. In 1993 they awarded the Nobel Prize in Chemistry to the developers of two tools, without which we could not now imagine doing modern molecular biology. Kary Mullis received it for inventing PCR (polymerase chain reaction) and shared it with Michael Smith, who pioneered the technique of site-specific mutagenesis.

Until Michael Smith came along, the methods available to geneticists for producing mutations were incredibly crude. The use of

X-ray irradiation, for which Muller was awarded the Nobel Prize in 1946, was still common as were a variety of chemical mutagens. However, those methods suffered from the disadvantage that the mutations produced were random, and elaborate screens or selective procedures were required to find the mutations of interest. Enter Michael Smith, who had learned how to make short synthetic oligonucleotides in Gobind Khorana's laboratory as a post-doctoral fellow. He realized in the 1970s, during a stint in Fred Sanger's laboratory, that with the sequence available for the DNA of a small virus, there was a possibility of using synthetic oligonucleotides to introduce specific mutations. This was accomplished using an *in vitro* methodology with a mutagenic primer and a DNA polymerase to elongate it. Site-specific mutagenesis was born. Together with Clyde Hutchison, Mark Zoller and many others, Michael Smith went on to develop the technique so that it could be used by researchers everywhere. This was truly a ground-breaking methodology that is now so widely used we have almost forgotten who invented it!

To appreciate the significance of site-specific mutagenesis we need merely look at the evolutionary process. During evolution random changes in DNA lead to random changes in the proteins encoded by that DNA and those that are beneficial eventually become selected. But natural evolution is a terribly slow process. We might wait for hundreds or thousands or even millions of years for truly advantageous mutations to arise. Thanks to the methodology developed by Michael Smith we can now hasten that process dramatically in the laboratory. We can introduce specific mutations into specific proteins at positions that we judge will make those proteins work better. In this way we can overcome the inherent limitations of slow natural processes to produce beneficial changes with great rapidity. Michael Smith was truly a pioneer whose insight, skill, and humanity make him a real hero of molecular biology.

— Richard J. Roberts
1993 Nobel Laureate in Physiology or Medicine
New England Biolabs, Beverly, MA
March 2004

PREFACE

by Eric Damer

One Saturday evening in 1994 I attended The Vancouver Institute, a free and longstanding public lecture series held on the University of British Columbia campus. The speaker was Professor Michael Smith, and his topic was his trip to Stockholm to receive the 1993 Nobel Prize for chemistry, awarded for his development of an important technique to manipulate DNA. I was curious to see a Nobel laureate for myself, especially since he was the first person from "my" university ever to have won the prestigious honour. I was immediately struck by his popularity — I sat in one of two overflow rooms since the main lecture hall of five hundred had filled early. Professor Smith proved to be a charismatic speaker with an easy-going and unpretentious manner. His pleasant and witty commentary was richly illustrated with elegant slides of the regal event in Stockholm. He closed his presentation by thanking his research associates and the university that had supported his work, and by offering a sincere thank you to the people of British Columbia for supporting the university and thus his career.

The audience in the main hall jumped to its feet and burst into applause. When Smith toured the overflow lecture halls, others stood and applauded enthusiastically. I marvelled at the response — the audience behaved like fans welcoming home a sports hero. Smith was no hockey player, yet all who had come that night to hear him responded overwhelmingly to the man and his achievements. To many people of the University of British Columbia and Vancouver, perhaps even the province and country, Professor Smith was indeed a hero.

When I was approached to help write a biography of Michael Smith I had a number of concerns. As is so often the case in historical research, information about Michael Smith's life was unevenly distributed with an abundance of archival records on his later years and very few on his early ones. No one had kept family records in the event that he would become famous, or so it seemed. Fortunately, there were several people whose memories could help fill those gaps. Because the biography would include recent history, I had to expect that I might stumble into current politics or recently settled or dormant controversies. My biggest concern, however, was that I did not simply want to write a hagiography. I knew many people admired Michael Smith as a scientist and as a person, but I wanted to peer beneath the rapidly developing myth to form my own view. I am grateful to have had the freedom to do so.

As I reviewed archival material and talked with his family, his partner Elizabeth, friends, colleagues, and students, I decided that the myth contained much truth. As a scientist and member of the academic community, Michael Smith was extremely well-regarded and well-liked for many good reasons. He was intelligent, creative, and ambitious in his work, but also modest and never ruthless. He had a well-earned reputation both inside and outside the academy for congeniality and conviviality despite a brusque manner and sharp-tongued sense of humour. In fact, I came to admire much about him as a person although I realized that he had his shortcomings and his failures. Many people told me that his ascension from rural English beginnings to the Canadian scientific elite "couldn't have happened to a nicer guy."

This biography is largely an account of Michael Smith's professional life, although personal elements are included to illustrate aspects of his character. His career in science might never have begun had the English school system not been sufficiently reformed by the 1930s to allow children from poor families to obtain an academic education. After completing his university studies, he still faced an uncertain academic future because of the nature of English society. In Vancouver, British Columbia, where he arrived almost by chance, Smith found a new home and began his academic life in earnest. For thirty years his laboratory at the University of British Columbia conducted scientific research of the highest quality.

Smith's ambition to excel as a scientist invariably led him into political issues, first in his own laboratory, then in his home institution, and later in Canadian science policy. He was, for example, part of the movement to "democratize" the University of British Columbia in the 1970s but part of the movement to "commercialize" the university fifteen years later. During the 1990s he played crucial but somewhat controversial roles in building new scientific institutions in British Columbia and Canada.

This account would be incomplete without some explanation of Smith's remarkable scientific work. This has been ably provided by molecular biologist Caroline Astell, a former student and colleague of Smith's, whose collaboration in this biography has been most fruitful and who was, in several key respects, also part of the story. She initiated this project and her participation ensured the cooperation of others.

It has been a pleasure and a privilege to have examined the life of Michael Smith, and to have been entrusted to write a biography that was fair and balanced while respecting his memory and the privacy of his family and closest friends. Perhaps in later years, with greater historical perspective, additional material may come to light that will add further dimensions to the story of this extraordinary man. Although the book is intended for the general reader interested in science, scientists, universities, British Columbia, the Nobel Prize, or Michael Smith himself, I believe that historians and scientists will also find something of value here.

One final prefatory remark seems appropriate. I have generally referred to the subject of this biography as "Mike" despite, I have been told, his preference for "Michael." I assume such familiarity because virtually everyone referred to him as Mike, particularly before his Nobel Prize. As the reader will soon learn, he had a very down-to-earth and informal personality that led to familiar appellations. But as a friend and colleague of his reminded me, he was no ordinary Mike.

— Eric Damer
March 2004

PREFACE

by Caroline Astell

I knew Michael Smith from 1964 until his untimely death in the fall of 2000. During that time our careers often intertwined and sometimes in very significant ways. Our first encounter took place at UBC while I was a graduate student. Mike sat on my magistral committee and later, as my doctoral supervisor, he provided a supportive environment in which to learn state-of-the-art biochemistry. Early on he instilled in his students and post-doctoral fellows self-motivation and a drive to succeed by granting them considerable freedom to pursue their particular research topic. Many students were not ready for such self-reliance and subsequently went their different ways — sometimes before completing their degrees — although to my knowledge all achieved considerable success in their chosen fields. Those of us who persevered with a career in scientific research surely would agree with Mike's dictum that the ultimate motivation is the thrill of discovery, whether in a simple project to improve the efficiency of linking oligonucleotides with cellulose paper or to sequence the genome of a newly emerged pathogen.

Throughout my career I have met many students who say they would rather interact with people than work in a lab. They believe that scientists toil for long hours alone in a smelly laboratory wearing a white lab coat — the stereotypical mad scientist. I learned in Mike's lab and others that this is not the case. I did wear a white lab coat, but I was surrounded by fellow lab workers who became a sort of second family, offering suggestions on my work, showing enthusiasm when experiments succeeded, and providing support when projects failed. Mike also initiated me into the often intense work schedule of the laboratory. Scientists spend many long hours in the lab but for those who really enjoy what they are doing, it is little different from pursuing a passion for mountain bike riding, gardening, or world travel.

Part of the excitement of being in Mike's laboratory was meeting high profile visiting scientists who talked with students and research fellows. In my case this included Gobind Khorana, Rich Roberts, Ben Hall, Edward Reich, Clyde Hutchison III, Chuck Dekker, Fritz Rottman, Peter Gilham, Roy Vagelos, and Fred Sanger, to name a few. Many of these scientists had won or would win the Nobel Prize for their achievements. Mike had an extensive network of first-rate scientific colleagues.

I left UBC for postdoctoral training at Rockefeller University in New York City. When I arrived I was in awe of the stature of many of the researchers. However, like Mike, many of the senior scientists at "Rocky U" were friendly and took a genuine interest in trainees. One of them, who recognized I was new, asked me where I was from, who my supervisor was, and what research I would do. Only later did I learn I had been talking with Christian deDuve, famous for his studies on fractionation of cellular organelles. My work in Mike's lab proved to be good preparation for my post-doctoral training.

In 1977 I returned to UBC where I again worked for several years in Mike's laboratory, partly to develop a new technique called site-directed mutagenesis but mostly to establish DNA sequencing and begin my own project: sequencing the genome of a small mammalian virus. I subsequently joined the Department of Biochemistry as a faculty colleague of Mike's where I continued my virus research until 2001. I am deeply grateful for four years of support by the British Columbia Health Care Research Fund and the continuous support of the Medical Research Council of Canada (now the Canadian Institutes of Health Research) for my research program.

In January 2002 I retired from UBC but soon found myself associated with the legacy of Michael Smith. When I returned from a holiday to New Zealand, I joined the Genome Sciences Centre at the British Columbia Cancer Agency as Projects Leader. This was the genomics centre that Mike co-founded with Victor Ling in late 1999 to implement genomics as a tool for cancer research. The GSC is currently Canada's largest genomics centre with a staff of over 140. Mike would have been proud to know that during the Sudden Acute Respiratory Syndrome (SARS) crisis in March/April 2003 "his" centre was the first to report the complete DNA sequence of the SARS coronavirus, allowing scientists around the world to begin devising diagnostic tests, methods to control the virus, and even a vaccine.

I first considered a biography of Mike in the summer of 2000 and contacted him with my idea. Our busy schedules prevented an immediate meeting, but we finally agreed to discuss the matter over lunch in October. Regrettably, his health failed rapidly towards the end of September and he passed away October 4. Although we never did discuss what a biography might include, I am confident he would have approved of *No Ordinary Mike* although, in his usual self-deprecating way, he would likely have said that he really had not accomplished that much.

I began writing my own account of Mike's life in November of 2000 but had little time because of commitments at UBC. When my appointment to the Cancer Agency ended my early retirement (almost before it started) I realized the book would not happen without assistance. I am indebted to historian Jean Barman for suggesting that I take on a co-author and for referring me to Eric Damer. Without Eric's hard work and more than considerable talent this book would never have been published. The result is an opportunity for the wider public to know about the career of a remarkable scientist and person.

— Caroline Astell
March 2004

I

A SHY LITTLE BOY FROM MARTON MOSS

January 1978. Persistent rain and blowing ocean winds made for a typically dreary winter month at the University of British Columbia. But the dedicated research team inside Michael Smith's cramped laboratory was oblivious to the bleak weather as they carefully tested the biochemistry professor's latest idea.

Mike bustled in and out of the small room, anxious to hear news about his team's progress. A year earlier, during a coffee break while on sabbatical in Cambridge, he had learned of a colleague's work inserting mutated strands of DNA into a host cell for replication. The procedure had limitations but Mike had seen in a flash how he could make improvements using his lab's specialty: a synthetic fragment of DNA. The first set of experimental results had been disappointing, but his team persevered, confident that the problem was with their laboratory procedures and not with their professor's brilliantly simple idea. Now, on their second attempt, having inserted a deliberately and precisely mutated fragment of

DNA into a cell for replication, Mike realized that they had devised a new method for manipulating the genetic basis of life. The implications were profound: new tools for diagnosing and treating genetic illness, the power to alter organisms in precise ways, and the capacity to produce useful biological products. The new technology would soon be hailed as the intellectual bombshell that triggered protein engineering, and its creator, an affable and unassuming British-born scientist, would garner accolades and awards as the father of site-directed mutagenesis. But at that very moment, Mike, knowing the significance of the results, wanted to celebrate in the customary way — with a bottle of inexpensive champagne. Grinning impishly, he popped the cork and poured the first round as his team celebrated their success.

The circumstances in which some people are born clearly suggest their future careers, but this certainly was not the case with Michael Smith. His place of birth in England was famous for tourism, not science, and he grew up in a rural hamlet known for its lettuce, hothouse tomatoes, and chrysanthemums. As the son of hardworking but poor market gardeners, it was highly improbable that young Michael would become a Nobel Prize-winning scientist able to influence government policy and public opinion. Sons of market gardeners in class-conscious England seldom acquired the necessary education or moved in the appropriate social circles. Mike was bright and had parents who valued education as a way to improve one's lot in life, but hard work and intelligence are rarely enough on their own to assure a rise to prominence. Mike was fortunate to be born just as the English school system was beginning to recognize and promote bright students regardless of family background.

Michael Smith was born April 26, 1932 in Blackpool, Lancashire, on the north-west coast of England sixty-four kilometres from Manchester. The seaside resort of one hundred thousand was a well-known and popular holiday destination for working men and women from the cotton mills, collieries, and factories of industrial England. During the 1930s, millions of visitors each year from across the country and abroad savoured the charms of Blackpool's beaches, ballrooms, piers, and pavilions, supporting the town's reputation as the busiest and brashest resort in the country. Gaudy entertainers, hawkers of cheap merchandise, and ice cream vendors eked out a living along the seaside promenade known as the Golden Mile, while the Blackpool Tower, an imitation of the Eiffel Tower, beckoned to hordes of pleasure seekers. Although the suburbs increasingly provided homes to service and light industry workers or retirees, Blackpool was more likely to produce famous entertainers than scientists.[1]

Mike was delivered by the District Nurse in the home of his maternal grandmother, Mary Martha Armstead, a few blocks inland from the seaside promenade at the heart of Blackpool's holiday district. Mary Martha operated one of the simple holiday boarding houses that accounted for Blackpool's working-class holiday appeal, and was among the "Blackpool landladies" who were a formidable social and economic presence in the city. Mary was a strong-willed character who ran her business with no thanks to her husband, who had earlier left her. Following Smith family tradition, Mike was baptized in the Church of the Holy Cross, a high Anglican church, a few days after his birth.[2]

Mike's mother, Mary Agnes — or Molly, as she was called — was an only child. She had done well in school, particularly in English and mathematics, and wanted to continue but was forced to leave in 1916 at the age of fourteen to work for her mother. The Blackpool holiday industries were notorious for employing young, seasonal workers, and Molly's mother wanted her to learn the boarding-house business at a young age. Yet for her generation, Molly had remained in school longer than most. Many young children from poor families left school at age twelve or thirteen, while children in Blackpool and other areas where child labour was popular often worked while attending school as a "half-timer." Molly was bright, hard-working, outspoken yet personable, and ambitious, and she resented leaving school to work for her mother. She subsequently learned bookkeeping in night school and found work in a tobacconist shop in South Shore, a suburb of Blackpool that was formerly a separate community.[3] Molly was determined to improve her lot in life, regardless of her mother's intentions.

One evening at a Blackpool dance hall Molly met Rowland Smith, a well-dressed, soft-spoken, and courteous young bachelor. He was not part of the Blackpool holiday industry, being the eldest son of a well-established local market gardener, who seemed to have opportunities to move up in the world. Molly soon found herself attracted to Rowland. Market gardening was a specialized and comparatively recent form of agriculture that did not become a major commercial enterprise in England until the seventeenth century. Before the railway age of the mid-1800s, fruit, vegetable, and flower gardens were largely confined to monasteries, wealthy estates, and the London suburbs. Because they required specialized horticultural knowledge, market gardeners often had more education and respect than other agricultural workers, especially if they owned profitable gardens or had wealthy patrons.[4]

To Molly, Rowland's family must have seemed eminently respectable. Edmund George Smith, Rowland's father, was an educated and qualified horticulturist who, as the family story went, had declined an opportunity to work in Kew Gardens, London's venerable botanical collection. He had even taught botany before leasing property in 1902, southeast of Blackpool in an area of peat bogs drained for agricultural use. Here, at Blowing Sands near the village of Great Marton Moss, Edmund grew lettuces, tomatoes, and flowers, especially chrysanthemums, which he took into Blackpool on his horse-drawn cart. Some of his crops grew in steam-heated glasshouses, a practice that expanded between 1900 and 1914 as a form of protection against early frosts, late winters, and the "black rain" from nearby Manchester. Although property was leased from Lord Clifton, the local landowner, the recent Market Gardeners' Compensation Act provided horticulturists with some financial protection for their fruit trees or more permanent plants should the landlord decide to evict his tenants. Market gardening was hard work, but after the economic slump of the early 1900s the gardens became increasingly lucrative as middle-class consumption rose. In this case, it supported Edmund, his wife (a descendant of Scotland's Rob Roy MacGregor, according to family history), their son, Rowland, and six more children in their own house. One of the daughters, Lucy, was Rose Queen in 1924.[5]

Rowland received a good basic education until age fourteen, achieving some academic success, but like Molly, he had to leave school to work in the family business. Outside of school, he distinguished himself locally as a runner. At the start of the First World War he went to Preston to enlist, lied about his age, and was sent to Ireland for air force training. Poor eyesight forced him to abandon hopes of becoming a pilot, and he spent the war as a quartermaster sergeant and truck driver for the Royal Flying Corps. He returned home hoping to pursue a career in textiles, but soon found it necessary to rejoin his father's gardening business.

Rowland Smith married Molly Armstead in 1926 despite hostility from her mother. Mary Martha never really approved of the marriage for some reason, and her relationship with Molly deteriorated over the years. Molly was, however, welcomed into the extended Smith family but she remained sensitive to any indication, real or imagined, that she was less than fully appreciated. She may well have felt self-conscious about her own family background. For the first few years, the young couple lived in a South Shore flat near the gardens and glasshouses of Marton Moss which was, until 1934, politically independent from Blackpool. Molly continued to work at the tobacconist shop while Rowland worked for his father. Money was tight, but Rowland risked a little each week betting on the greyhound races in the hope of winning enough money to pay their rent. He often did win, but never large amounts.

By the time Mike was born in the spring of 1932, the family had bought a plain, brick house on Stanley Road (renamed Squire's Gate Lane) close to Rowland's father's home and gardens. Edmund, now a widower, had taken an early retirement, so Molly left her job to help manage the gardens. Because Edmund was not a particularly generous employer, Rowland leased additional property to grow his own flowers and vegetables. His hobby of raising budgerigars also added to the family income when his birds won local or national prizes and sold at a good price. Even so, the

Smiths had little extra money during the world-wide Depression of the 1930s, leaving Mike without many toys or treats. After the modest prosperity and population growth of the 1920s, the Depression had returned Marton Moss residents to a familiar, harsh lifestyle. Market gardeners generally fared better over-all than grain-growers but worked long, labour-intensive hours planting, weeding, and picking crops by hand, maintaining drainage ditches and glasshouses, and sterilizing soil every winter with steam produced in converted railway boilers. The Smiths, like others, made small gains through hard manual work and quiet determination. Although a little money could be spared to provide Mike with the occasional luxury — perhaps a trip to the circus or a Gilbert and Sullivan operetta at Christmas — it was a time of austerity.

Mike was too young to understand fully his family's hardships during the Depression and had a reasonably happy childhood in Marton Moss. From an early age, the cheerful, red-haired little boy exhibited a number of distinguishing personality traits. He loved to play games and share jokes with his friends, but he was sensitive about whether others wanted his company and often hesitated to join group activities until reassured that he was welcome. Although friendships were important to him, he liked winning contests and sometimes became upset when he lost. To balance his needs for social approval and personal achievement, he learned to be persistent without appearing ambitious. Mike's good-natured sociability was encouraged by his parents who were very fond of their son, although Rowland spent most of his time in the gardens and left Molly with the child-rearing duties. Mike grew up learning that mothers were supposed to stay home and care for their children, but Molly also managed the family income and did all the cooking, washing, and cleaning in addition to her gardening chores, instilling in her son a strong work ethic and the belief that women were capable of working inside and outside the home. As an independent-minded youngster, Mike was uncomfortable in formal or authoritarian settings like his Sunday School classes at St. Nicholas Church, where Molly occasionally went for Sunday services. He would later remember winning an award for attendance, but nothing else.[6]

Young Mike also absorbed aspects of the colourful Marton Moss culture. For centuries, the "mere-town" was isolated from Blackpool to the north and Lytham St. Anne's to the south by treacherous bog, sand dunes, and dense forest, oblivious to the developing tourist industry nearby. In 1873, a Blackpool newspaper reporter described the area as "intensely rural in its aspect" with the inhabitants having "a certain primitive look about them." The local dialect even contained traces of Old English and Norse. Intermarriage among a handful of old families made for a close-knit community that relied on colourful nicknames to distinguish one person from the next. By the 1930s, a century had passed since the Reverend William Thornber dismissed Marton Moss as "little less than a scene of moral destitution," but common-law marriages, rowdy sports, embarrassing and sometimes malicious pranks, and heavy drinking were still part of local life. Few houses had plumbing although some street lamps were fuelled with sewer gas, while windmills, white-washed brick cottages, thatched roofs, and horse-drawn carts remained common. The village blacksmiths still forged horseshoes from hot, glowing iron and repaired the popular wooden-soled boots (clogs), incidentally providing free entertainment for Mike and his friends. Needless to say, few people completed high school and almost no one went on to university. Mike would forever be known for the down-to-earth mannerisms, simple tastes, and earthy sense of humour that characterized people from the Moss.[7]

Mike was seven years of age in 1939 when Britain entered the Second World War, the same year his brother Robin was born. Sometimes at night the Smiths could hear enemy bombs falling in distant Liverpool and occasionally the Nazi bombers would target the nearby aircraft factory that produced Wellington bombers. Mike was home alone one night with Robin when bombs fell on either side of the house, possibly because the glasshouses were highly visible as targets. The terror of the bombings began for Mike what became a life-long opposition to war. Fortunately for local residents, Marton Moss was generally safe from air raids and a few families billeted children from Manchester and London who had been evacuated for their safety as part of "Operation Pied Piper." Blackpool itself prospered as a training centre for Royal Air Force personnel and as one of the few seaside holiday resorts operating during the war. Molly and Rowland, encouraged by government agriculture policy, supported the war effort by growing food crops instead of flowers. Mike, observing the long, hard days worked for low pay, decided that this was not the life he wanted to lead.

Although Mike's childhood was poor and unglamorous it would be wrong to suggest that the Smiths were typically "working-class," a label that applied to some seventy-eight percent of the English population in 1931. It was true that Rowland and Molly worked at hard manual labour for low pay and were far from wealthy. Rowland also enjoyed betting on greyhound races, a popular working-class sport, and his hobby of breeding budgerigars was, like pigeon or canary breeding, a widespread working-class activity. Moreover, because he could not afford to join a middle-class sports club, his past success as a competitive runner granted little prestige. Yet Molly and Rowland, coming from entrepreneurial families that owned property and ran successful small businesses, embodied many middle-class values. They had done well in their years of schooling, unlike the majority of working people of the same generation, and they were keen to see their sons succeed in school. Molly's specialized training in bookkeeping put her in the clerical ranks, and the special horticultural knowledge Rowland had learned put him above unskilled farm hands elsewhere. They were literate, numerate, and aware of the potential for education to provide opportunities for a more prosperous career.[8]

The Smiths displayed other distinctions and ambitions, both for themselves and their children. Rowland served in the Royal Flying Corps during the Great War rather than the army and always presented himself as a quiet and courteous country gentleman, well-dressed and fair-minded. Sometime around 1940 he acquired one of the few cars in the village, a twelve-horsepower Rover he used for transporting flowers and produce. Although Rowland rarely attended services at St. Nicholas, both he and Molly identified with the Church of England, the established church of middle-class respectability. They took modest leadership roles in their local community when they could. Molly, for example, helped to organize local theatrical productions and Rowland officiated at school sports days, later acting in local plays. Perhaps most significantly, the Smiths aspired to homeownership for most of their working lives and were ultimately successful. In these respects, they had middle-class aspirations and sought opportunities to improve their fortunes and circumstances, protecting whatever wealth already acquired.

Molly and Rowland imparted their ambitions to their son, who needed little encouragement to choose a life without hard physical labour. Educational success seemed like the best — perhaps the only — way for a bright but poor boy to attain a better life, but the only school available to Mike in the beginning was St. Nicholas Church of England School, formerly Marton Moss Church of England School, but known locally as Clog College. It was a large brick building with extensive grounds that also served as a centre for social events, including whist drives (Molly was a keen card player) and snooker games by members of the local Men's Institute. Formerly an "all ages" school, it divided into an infants and junior school in 1939 catering to ages five through eleven. Mike was fortunate that he attended St. Nicholas when he did. Had he gone there a few years earlier, it is unlikely that he would have been given the chance to succeed.

St. Nicholas was a poor, semi-rural school opened in 1873 by the Church of England at a time when Lancashire social reformers threatened to open a secular or non-sectarian state school in the area. Students at Marton Moss School, as it was known in the early days, often missed classes due to illness or stormy weather, and parents frequently kept their children home to help harvest vegetable crops. It was not a strong academic school, but by the time Mike arrived educational reforms across England in the preceding decades had raised the school-leaving age, prohibited early leaving, abolished the half-time system, and set minimum qualifications and competitive wages for teachers. Reforms paused during the height of the Depression, but resumed by 1933 and provided improved opportunities for disadvantaged students to obtain an academic education. Although church-sponsored schools had less access to public funds, Marton Moss School complied with the new standards in education by enlarging in the mid-1920s and by hiring qualified teachers. In 1929, for example, Percy Patrick Hall, a certified teacher and Labour Party supporter, began a distinguished career in education and politics at Marton Moss School. As a student at the renamed St. Nicholas School, Mike received an education that gave him the grounding for future studies — if an opportunity for those appeared. Certainly, Mike always placed at or near the top of his class.[9]

Yet a good education at St. Nicholas was far from adequate if Mike wished to climb the educational system to its top. Elementary education in England had originally been designed to provide poor or working-class children with a minimal and terminal education before entering the workforce. Wealthier parents tended to send their children to private schools. For Mike truly to raise his social and economic prospects, he would need to find his way into a grammar school, as the academic secondary schools were called. Grammar schools traditionally were private and exclusive, requiring high to exorbitant fees. (Because they were constituted nonprofit charities, many private schools, including the elite boarding schools, were called "public schools.") Even the new state-supported grammar schools established after 1902, a concession to middle-class demands and a growing need for clerical workers, charged fees. Poor children could only dream of winning a scholarship from the Local Education Authority (LEA) to attend a state-supported school, but until the 1920s they rarely did. Subsequent school reforms spurred by the increasingly popular and now middle-class Labour Party (which promised "secondary education for all") helped to expand the number of secondary schools and scholarships, challenging what some considered an elitist social system buttressed by a small number of highly exclusive private grammar schools.[10]

As fortune would have it, just at the time Mike was distinguishing himself as a bright young student, opportunities to attend a grammar school were also improving. Despite interruptions during the Depression, the Blackpool Education Authority that now supervised Marton Moss was reforming its schools along the lines suggested by the Hadow Report (1926), converting its "senior," "central," or "intermediate" elementary schools into secondary schools with higher standards and the administrative status of the grammar schools. By the late 1930s, legislators had responded to public pressure to provide free secondary education for all students by raising the school-leaving age to fifteen (to begin in 1939) and expanding scholarships to grammar schools. The outbreak of the Second World War stalled school reforms but the social and political mood to change the educational system under the war-time coalition Conservative/Labour government was very strong.[11] Yet until those ideas became law with the Butler Act of 1944, grammar schools still charged fees and accepted only a minority of students on scholarship.

In the spring of 1943, at age eleven, Mike faced the critical juncture that would decide whether he would proceed to a free place in a grammar school (by means of a scholarship) or complete a "post-primary" non-academic elementary education. His fate depended on his performance on the "eleven-plus" exam. If he did well, he would win a scholarship to a grammar school; if he did not, then there would be no secondary education. (Students writing the eleven-plus exam after the 1944 Act had to score well to attend a state grammar school; they could no longer attend simply by paying fees.) However, even with these new possibilities in education, class background played an important part in the choices of schools. Most of Mike's friends would not go to a grammar school but to Palatine School for a few years of technical education or to Highfield Senior School for a few years of "rural science," both schools still under the elementary code. Molly certainly wanted her bright son to do well in the most challenging program; she felt he was just the sort of boy the school reforms were meant to help. Mike also wanted to do well on his eleven-plus, knowing that his family could not afford school fees and wishing to have an easier life than his parents who toiled from dawn to dusk for little pay and no vacations. Mike's success would depend on how well he could handle the pressure of writing the exam and whether his previous years of schooling had been adequate.

The eleven-plus exam varied from one location to another, but it typically tested a student's knowledge of English, mathematics, and "general knowledge." The rise of psychology and intelligence testing put special emphasis on verbal reasoning skills, and proponents claimed that it was a simple and accurate way to measure innate intellectual ability. It would be some time before social scientists would question whether the exam unfairly discriminated against students from poor backgrounds, although intelligence testing had been used to show that many fee-paying grammar school students had mediocre ability. The exam had also become very competitive, encouraging streaming in elementary schools in Blackpool and across the country and creating considerable stress for many children. Much was at stake for Mike and his parents, but despite any handicaps that may have existed because of his background, he did very well in the crucial eleven-plus. In fact, Molly heard that her son's test result in mathematics was the highest in Blackpool.[12] As a top scholarship winner, Mike could choose his grammar school.

The choice was simple, at least for Mike. Cousin Dennis, two years older, was at Blackpool Grammar School so it seemed a natural destination. Mike wanted to be with people he knew and liked, but his mother had a different plan. She wanted her son to attend Arnold School for Boys, probably the best academic school in the area and easily accessible by bicycle from their house on Stanley Road. At first Mike refused. He did not care for Arnold School at all and worried that his current friends would tease him if he attended. They thought that Arnold School students were snobs, and Mike did not want to risk losing his friends. He begged his mother not to send him there, but Molly, a woman of iron determination who was accustomed to managing her family regardless of opposition, would have none of it and insisted her son enrol.

▲ Michael Smith at approximately one year of age.
(COURTESY UBC ARCHIVES)

▲ Mike's parents, Rowland and Molly Smith in Manchester, 1953. (COURTESY DENNIS SMITH)

▲ Stanley Road (now Squire's Gate Lane) in the 1940s, near Mike's childhood home. (COURTESY BLACKPOOL CIVIC TRUST)

▲ Glasshouses similar to those of Mike's father near Marton Moss, 1930s. (COURTESY BLACKPOOL CIVIC TRUST)

▲ St. Nicholas Church where Mike occasionally attended Sunday school classes. Marton Moss, late 1930s. (COURTESY BLACKPOOL CIVIC TRUST)

▲ Mike and his classmates worked in the gardens at St. Nicholas School, 1930s. (COURTESY BLACKPOOL CIVIC TRUST)

▲ An early view of Mike's elementary school, St. Nicholas, before expansion in the 1920s. (COURTESY BLACKPOOL CIVIC TRUST)

▲ Arnold School: Mike's grammar school from 1943 to 1950.
(COURTESY ARNOLD SCHOOL ARCHIVES)

▲ Arnold School students in a display of physical training, 1936. An interesting comparison with the students of St. Nicholas School.
(COURTESY ARNOLD SCHOOL ARCHIVES)

▲ Mike as a teenager on an excursion into the English countryside. (COURTESY UBC ARCHIVES)

▲ Mike's twenty-first birthday party, 1953. Mike is front row, second from left. Brother Robin is on Mike's left, cousin Dennis is second row, far right. (COURTESY DENNIS SMITH)

▲ The University of Manchester Honours Chemistry Class of 1953. Mike is in the front row, fourth from the right.
(COURTESY UNIVERSITY OF MANCHESTER ARCHIVES)

▲ Khorana's laboratory members at the B.C. Research Council in the late 1950s. Mike is seated on right, next to Har Gobind Khorana. Behind Mike in lab jackets are John Moffat (left) and Gordon Tener (right).
(COURTESY UBC ARCHIVES)

Mike approached Arnold School with great trepidation. Arnold had been established as a private, proprietary school in 1896 by Frank Pennington, its principal shareholder and first Headmaster. Pennington's school was not the first private school in Blackpool to attract sons of the prospering middle-class, but it had done very well in building a day and boarding clientele. It began in a local Masonic Lodge, but in 1906 it moved to grander buildings on Lytham Road, South Shore, and by 1939, the school had 286 students of whom about two-thirds were day-students. The school required an entrance examination and had a good academic reputation before and after its first state inspection in 1917, when Pennington was praised for his leadership and Arnold School was certified as "efficient."[13]

Boys like Mike were not usually found at Arnold School. Although the school was not a traditional gatekeeper to high office as were such famous boarding schools as Eton, Harrow, Rugby, or Charterhouse, it was nonetheless exclusive. Part of the exclusivity was cultural, as the school emulated these leading public schools in its subscription to Anglicanism, school uniform, authoritarian management and student hierarchy, militarism, sport, elocution, and manners — character building, as it was called. Exclusivity was also ensured by charging higher fees than state schools, although Arnold's were not as high as many other private schools. Still, they were high enough to keep out students like Mike.[14] Yet once again circumstance was in his favour, for shortly before Mike was of the age to enrol, Frank Pennington reorganized his school as a non-profit charity. This meant that Arnold was now a state-oriented "public" school eligible for government financial support during the economically lean 1930s, and it began to receive the so-called "direct grant." When Arnold had been a private school supported by fees and profits from the award-winning school farm, it had not been compelled to accept scholarship students. Now, after reorganization, the school took the further step of opening its doors to local scholarship holders. Molly could send her son to Arnold School as a day boy, providing an experience that would have a great influence on Mike.

Mike nervously entered an unfamiliar and intimidating world, an experience he shared with other less affluent boys elsewhere who found themselves out of place in a private school. Few boys from St. Nicholas School went on to further academic education, and Mike was one of two boys from the Moss to register at Arnold School that year. As far as anyone knew, they were the first. It was a world of "old boys," school traditions and rules, strict headmasters, debating societies, sport, school prayers and songs, uniforms for all occasions, cadets, elocution, and Saturday night dances with students from Arnold High School for Girls. Character development of the "leadership" sort was almost as important as academics. Most of the teachers had university degrees at a time when few had them, and many had whimsical nicknames such as Thistle, Bumps, Knicknacks, Budda, and Beaky. Even sport, a vital aspect of the public school tradition, had its own particular flavour that was alien to young Mike who was small and not particularly well-coordinated. He had little talent in rugby, cricket, football, grass hockey, or athletics (track and field), leaving him unpopular, socially awkward, and insecure. His protruding front teeth, about which he was already sensitive, caused Mike additional anxiety when students teased him about his overbite and his rural accent. He worried that his schoolmates would discover that he had only one pair of grey flannel trousers that he kept under his mattress at night to make them appear pressed in the morning. For a boy who valued his friends, Arnold School was a very unwelcoming place.[15]

Curiously enough, in later life the thing Mike most recalled about Arnold School was the atrocious food served to him. He hated the lunches of blackened boiled potatoes and suet puddings, and once vomited on the prefect who made him finish his unappetizing meal. Mike thus joined the likes of George Orwell who, at his preparatory school during the First World War, complained that his porridge "contained more lumps, hairs and unexplained black things than one would have thought possible" and was served in bowls encrusted with "accumulations of sour porridge which could be flaked off in long strips." At least Mike avoided the tripe that the Headmaster had earlier secured for his school that was served cold with vinegar or hot with onions. Wartime shortages meant that food was rationed, vegetables were grown on the school grounds, and treats in the snack shop were curtailed. Meals were only the most obvious aspect of the school that Mike disliked. There was also the hierarchical discipline and the sense of always being ordered about. Once Mike received a blow from "the slipper" on a "rather tender part of [his] anatomy" for the simple offence of throwing a paper dart. Although as an adult Mike did not tell horrendous stories about physical and mental abuse (as have disgruntled old boys at other schools), discomfort during the mid-day meal reinforced the anxiety that accompanied Mike's experience at Arnold School.

Since Mike was at Arnold during the war years, he and other students helped with the national war effort. The students collected money and helped local farmers bring in their crops, a small but symbolic contribution. Mike of course was no stranger to rural life, but he later recalled how unpleasant it was wading knee-deep in the mud picking potatoes. To prepare for air raids, students practised wearing gas masks, dug shelters in the playing fields, and criss-crossed windows with adhesive tape to prevent splintering. Even with nightly black-outs, a dozen or so bombs fell on the cricket field during the course of the war. Arnold School hosted students from Manchester Central High School earlier in the war as part of the national evacuation campaign, but most of them had returned home by 1943.

Yet even with all these war-time activities, Arnold School continued to do what it did best, which was to emphasize its academic program, and here, at least, Mike flourished. At first, like other working-class children who won scholarships to grammar schools, Mike was relegated to a less academically demanding stream. Unlike many others, however, he persevered in the new culture and distinguished himself academically by ranking third in his class of twenty-nine in the first term and first in the second term. He was soon moved to the top academic stream for those deemed fit to enter a profession or university. Mike developed strong, self-disciplined study habits through homework assigned in three subjects every night that was graded for frequent reports to parents. With his strong powers of concentration, Mike studied hard and regularly placed near the top of his class, giving his mother good reason to boast about her son's achievements. Although Mike's "house," the Romans, never did particularly well in sports, they were often at the top of the fortnightly marks competition.[16]

Two people seemed particularly interested in helping the shy little boy from Marton Moss. Mrs. Holdgate, the wife of the new Headmaster, impressed Mike with her kindness. (She had earlier impressed the school's governing council, helping her husband, Frank Holdgate, become "the Boss.") The other person to take a personal interest in Mike was the school's science teacher, Sidney Law. Mike may have appeared shy but all he needed to overcome his shyness was to feel accepted by others. Mrs. Holdgate and Sidney Law helped him to feel a little more welcome, and Law introduced Mike to a new academic interest.

Sidney Law came to Arnold School first in 1927 with a science degree from London University to teach physics and chemistry. Students nicknamed him "Quiffy" because of his shock of hair, but this appellation was short-lived. He appeared to some a "figure of authority who did not suffer fools gladly," but students also found him patient and kind, attributes lacking in some teaching masters. He left Arnold in 1934 to teach science elsewhere but returned in 1944. Law first taught Mike in a general science class that included a unit on earthworm anatomy, which would, curiously enough, prove to be Mike's only biology lesson ever. Science was new to Mike. As a youngster, he had never been particularly interested in the subject. He had never collected butterflies, solved difficult mathematics puzzles, or formulated his own theories on the popular physics of the day. Over the next few years, with Law's encouragement, this changed. Mike never achieved top marks in language-based courses, but he excelled in mathematics, physics, and especially chemistry. Perhaps, like other budding scientists of the era, Mike found science to be something of a refuge from strict headmasters and bullying peers.[17]

Moreover, Mike's new academic interest came at a time when

post-war England had a renewed interest in science after confronting the military power of German technology. The First World War revealed that British science lagged behind that of other countries, and the nation responded quickly during the war with university-based research, especially in chemistry. Preparations for war-related research were begun before the Second World War, especially in physics, and the war itself stimulated innovative developments in such technologies as radar and atomic energy. Following the war, government and industry cooperated to provide more opportunities for bright students to acquire university-level education of industrial value. Arnold School responded to the new post-war developments by strengthening its sixth and highest form (grade), expanding its science classes, and by encouraging capable students to apply for university admission.[18] With Sidney Law's friendly encouragement, Mike applied his academic talents to chemistry.

Mike may have known little of government or educational policy of the time that had given him his opportunities, but he did know he had found his passion. He did well in his science studies at school and continued them at home. Mike's parents encouraged his interest in chemistry by installing a gas line to a shed in the back yard so that their son could operate a bunsen burner without endangering the house. They may also have provided him with the money to purchase inexpensive chemicals so he could try experiments at home, although sometimes his Roman candles failed to shoot fiery balls and his rockets often remained on the ground. His experiments at Arnold also went astray on occasion. Once, an ambitious experiment with burning magnesium wire left a scar on the laboratory floor.[19] Undeterred by such minor setbacks, Mike wrote the junior matriculation exam at age fifteen and moved into the higher forms where he concentrated on mathematics, physics, and chemistry, both organic and inorganic.

Outside of Arnold's formal academic curriculum, Mike slowly learned some of the values of his new social setting. Sidney Law, for example, suggested that Mike read a better quality newspaper than the local paper popular with his parents and other Mossogs. Mike subsequently began reading the *Manchester Guardian*. Improving Mike's character was another challenge. Sport was not particularly helpful because of his poor coordination, although he was required to practise several times a week and march with the cadets. Teachers reported only "fair" ability in speech and physical training. A partial solution to Mike's character development came unexpectedly in response to his overbite. The Headmaster sent Mike to a local dentist at the end of his first term to consider corrective measures, but instead of fixing his teeth, the dentist introduced Mike to the leader of the 16th Blackpool Scout Group. Scouting was socially compatible with Arnold School and instantly appealing to Mike, who soon joined the troop.[20]

Mike took to Boy Scouts with great enthusiasm although most boys at Arnold were uninterested. During the war, the group met above Longstaff's Boot Shop, South Shore, after the Royal Air Force appropriated their former meeting hall. Like other Scouts in England, the members of 16th Blackpool watched for fires following air raids (they were too young to take a more active role), collected paper, and helped erect air-raid shelters for elderly residents. The group leader, Mr. Barnes, was also an accomplished outdoorsman keen to take his boys hiking and camping. One popular destination was the Lake District a few hours drive north of Blackpool. England is not a particularly mountainous country, but the Lake District, with its low peaks, crags, and lake-filled valleys is an area of natural beauty that has inspired tourists and poets alike. Mike hiked, camped, fished, and climbed the rock bluffs of the area, developing a love for the outdoors that stayed with him throughout his life. He spent a few of his summers working at the Great Towers Scout Adventure Camp, a 250-acre site on the shore of Lake Windermere. Following the war in 1947, Mike represented his group at the Jamboree de la Paix, the first post-war international jamboree in France that attracted forty thousand Scouts.[21]

Scouts meant camping, hiking, and woodcraft, but it also reinforced certain social values. The movement's founder, Lord Robert Baden-Powell, was a decorated and celebrated military officer from a professional and well-connected English family. He had enjoyed an education at the prestigious Charterhouse School and briefly studied at Oxford University. Before his death in 1941, Baden-Powell's popular youth movement had come to represent an informal yet formidable educational force intent on moulding character. Scouting combined camaraderie and the excitement of outdoor adventure with the spirit of Victorian public schools and a paternalistic yet compassionate social conservativism. As a religious but non-denominational movement appealing to middle-class youths, Scouting was an appropriate complement to Mike's life at Arnold School by providing physical recreation with new friends who suited his new school culture.[22]

Over the years, Mike found Arnold School less intimidating. His cheerful, unassuming, and easygoing disposition earned him some acceptance from peers and teachers. As he would for the rest of his life, he made great efforts to adjust to new and uncomfortable circumstances. He rose to the rank of sergeant in the cadets and determinedly practised with the rugby team to earn some recognition in sports. As a school monitor in 1949, Mike helped to enforce school rules and protocol during lunches (which were a little better after the war), although he never did care for Arnold's formal and authoritarian organization. He even endeared himself to a few fellow students whom he helped with their homework. In subtle ways, however, he felt himself a social outsider and never attempted to acquire the polite manners and refined accent that would hide his humble origins; no one would ever think he had become a snob.

In 1950, Mike wrote his senior exams, and again did well. His school report in chemistry contained "nothing but praise," although the previous year he had been admonished for spoiling his work with "careless, illiterate, childish mistakes." He won a Blackpool Education Committee Major Scholarship that covered the low tuition costs for a university degree and, more importantly, provided £160 per year for living expenses.[23] Although Arnold School students often won local and national scholarships, for Mike the financial support was crucial for him to continue his education. Mike's parents were also pleased. Molly especially had encouraged and reassured her son over the previous seven years, and she knew that she and Rowland were unable to provide the tuition and living costs for a university education.

But which university would be best for Mike? The universities with the highest prestige for the brightest scholars were Oxford and Cambridge. Oxford had enjoyed a first-rate chemistry program since before the war and Cambridge had recently joined the forefront of the field with its revitalized program.[24] Arnold School employed teaching masters with Cambridge degrees, and a number of Arnold graduates in the past had gone to Cambridge. Yet these universities still clung to their medieval roots and traditions in ways that worked against Mike. They had long cultivated close ties with established wealth and power, and still tended to serve an exclusive student clientele. Mike certainly lacked social privilege. As state grants increased during the twentieth century, Oxford and Cambridge married merit to social connection but retained what to Mike was a formidable barrier.[25] Oxford until 1960 and Cambridge until 1961 also required an ancient language, Greek or Latin, as an entrance requirement for science students. Mike had studied Latin at Arnold School, but for some reason he did poorly and left the subject to concentrate on science. French, the other language taught at Arnold, or scripture studies were no replacements. Without Latin, Oxbridge was not an option.

Fortunately, nearby University of Manchester had a good chemistry program of its own. In fact, it was excellent. Oxford's standing in chemistry had largely been built on expertise poached from Manchester after the First World War, and Cambridge revitalized its chemistry program with expertise it poached from Manchester during the Second World War.[26] Several Arnold teaching masters held degrees from Manchester while others, including Sidney Law, held degrees from other civic universities. Many Arnold graduates had gone to Manchester in the past and the school had long ties with the university. Mike decided he would enrol at Manchester, the destination of several of his peers that year, and study for a career in the growing field of industrial chemistry.[27]

The University of Manchester was one of England's well-regarded civic or "redbrick" universities that had catered largely to a regional, middle-class clientele since its origins in 1851 as Owens College. As the University of Manchester after 1903, the institution maintained its leadership in chemistry research and education and did important studies during the First World War on such chemicals as toluene (for the explosive TNT), acetone (for the propellant cordite), and lethal gasses. Although some of the research was theoretical, the university was mostly practical and utilitarian as dictated by the industrialists and, increasingly, government representatives who dominated the hierarchical governing councils. By the 1950s, Manchester was also renowned for its medical school, historians and Hebraists, the world's first electronically stored computer program, and three Nobel prizes — to mention a few achievements. Although the post-World War II student enrolment boom had subsided by the time Mike graduated from Arnold, the university had grown considerably thanks to increased state funding.[28]

Academically, Mike had made a good educational choice. But as he soon discovered, he was going to one of England's bleakest university cities, subject to "dense fogs, industrial illth, and a bronchitic sub-climate that depressed most newcomers." Buildings were black from nearly a century of coal smoke and still bore scars from the war. It was austere and thrifty, and not accustomed to the pomp of the 1951 centenary celebrations that brought George VI's consort to the campus. Students tended to be sober, serious, and deferential to authority, with little clubbing and pubbing outside of the annual "Rag Week" carnival. Many were from modest, middle-class homes and worked hard with local scholarships to help improve their career prospects.

The University of Manchester presented a special problem to students like Mike who came from out of town. Many students lived at home while only about one-quarter found accommodation on campus. The rest, nearly forty per cent of the student body, had to find lodgings, or "digs" as they were called, in local boarding houses or in shared flats with other students.[29] Mike's "digs" the first year were not so good. His boarding house room was not very comfortable and he told his cousin that he survived by eating tins of baked beans. The following year he found much better accommodation with the Elton-Jones family near the pleasant suburb of Sale. Mr. Elton-Jones sold pressure cookers, a new and exciting appliance, in the Lake District. He and his wife had a large house and three teenage daughters, but boarded students to help with expenses. The family grew quite fond of their new boarder, providing him with good meals and a lively social atmosphere that was so important to him. Mike and the other male boarders especially enjoyed mingling with the teenage daughters, although the university students imposed strict homework rules on themselves during the evenings. Mike enjoyed his new "digs" and stayed for most of his student years, making long-lasting friendships with some of the other boarders.

Manchester offered various science degrees, but Mike's strong interest in chemistry led him to the Honours program. Many students took a three-year Honours degree, part of a growing trend towards specialization, and the class that began in 1950 numbered sixty students. Only two of these were women, confirming that the university and particularly science were still male preserves. Students only occasionally saw the department's two professors (one of organic chemistry, one of physical chemistry) who held almost total power over the department and were members of the university's academic Senate. Instead, readers and lecturers taught Mike and his colleagues the core subjects of physical, organic, and inorganic chemistry, along with mathematics and physics. In addition, demonstrators and graduate students helped provide some twenty hours of practical laboratory work each week. Saturday mornings could be used to compensate for laboratory work missed because of student inattention or continuing post-war power failures and shortages of supplies. Mid-term exams took place in the fall and spring, and the most important final exam in June.[30] Because many chemistry publications were in German, Mike and other students studied to pass a language proficiency test.

Like the university in general, the Department of Chemistry had a formal air. Mike was called "Smith," academic staff were called "Sir," and tweed jackets and neckties constituted proper attire when not in the laboratory. Mike did not much care for these formalities, but at least he no longer had to wear a school uniform. Staff and student social events were infrequent, but the occasional Christmas party, cricket match, or feast of Lancashire hot-pot and beer provided welcome interruptions to a heavy study load. Mike did not find his lectures always to be of high quality. In part this was because of the emphasis on research publications to advance academic careers, and students had not yet learned to complain openly about the quality of their education. Many students after 1953 read the novel *Lucky Jim* and sympathized with the character Dixon, a restless young history lecturer from the lower middle-class frustrated by the stodgy and conservative world of the college establishment. Mike and his peers enjoyed or endured their lectures, comforted by the fact that they were learning chemistry in a well-respected department and with the knowledge that their career prospects were excellent.[31]

As a student at Manchester, Mike was good but not outstanding. He rarely scored top marks in his exams and even failed two in his second year and one in his third. Fortunately he never failed a June exam, the one that really counted. Some of Mike's instructors had mixed opinions of him, noting in his first year that he was "quite intelligent" but "rather a rough diamond." The second year, he appeared as "cheerful, interested" but had only "moderate ability" in physical and inorganic chemistry. In the third year, Mike was again judged to be "cheerful, but [with] only moderate ability" in inorganic chemistry. He was seen as "very intelligent, [but] not so interested in organic chemistry." Other students thought he worked hard and had a broad and independent view of the subject, but this did not always win him favour with his teachers. Once, during a laboratory class, Mike was expected to produce a standard derivative compound to identify a sample but unaccountably (and with a small grin, his friends thought) he produced a dye — much to the annoyance of the laboratory supervisor. Still, his grades were generally good, especially considering the difficulty of the program. Each year several students transferred to the general program (only forty-four of the original sixty students entered the third year) but Mike was never in danger of joining them.[32]

True to his nature, Mike sought out new friendships while at university and was relieved to find classmates who cared more about science than his accent and awkwardness with a cricket bat. He made many new friends who shared with him various outdoor pastimes, including weekend hiking and rock climbing trips to the Lake District or Derbyshire Dales. Much to his tastes were the Saturday evening pub-nights, as well as excursions on the back of a friend's 1937 Triumph motorcycle that provided additional inexpensive (and sometimes dangerous) entertainment. Mike discovered he liked beer and would join his friends for a friendly pint. He had his first travel adventure with a couple of friends in the summer of 1952, hitch-hiking through France to the Mediterranean and returning to Paris where they discovered the exotic and glamorous entertainment of the local cabarets, a marked contrast from the pubs back home. In 1953, Mike joined the crowds to watch the sensational 1953 Football Association Cup final between Blackpool and Bolton that Mike attended thanks to a free ticket from one of his father's friends. (Blackpool won 4–3, though trailing 1–3 with only twenty minutes to go.) Although somewhat shy at first, Mike was extremely sociable, and his pleasant ways won him many new, lifelong friends who valued not only his dependability and honesty but also his sense of humour and love for life.[33]

One new and enduring pleasure introduced to Mike while a student was classical music. He was soon in the habit of buying inexpensive seats for about twelve pence to attend the well-known Hallé Orchestra concerts conducted by Sir John Barbirolli, joining university lecturers and the "self-made men of Manchester" who were proud of their cultural institutions. Once, after a climbing trip, Mike and a few friends marched directly into the Free Trade Hall for a concert. As they walked up to the inexpensive seats located high above the others, a woman noticed the climbing rope and remarked, "It isn't that high a climb!" On another occasion, during Rag Week, Mike and a few of his more mischievous friends strung a banner across the stage during an opera to advertise their charity fundraising. Somehow they managed not to interrupt the performance and yet convinced the audience to contribute to the charity.

Mike also read widely in literature and other subjects. He often told his friends about the most recent novel he had read — Cervantes' *Don Quixote,* for example, or lighter American fiction — and he often initiated debates over articles he read in the *Guardian* newspaper. He expressed concern for social welfare and usually promoted a left-wing perspective, perhaps under the influence of an outspoken, leftist friend.

Although Mike's new university friends saw him as an affable and intelligent student colleague, he inwardly felt handicapped by a lack of self-confidence. With his friends he could be bold, outgoing, and cheeky but in other circumstances he was timid and withdrawn. He was especially shy with women, although he had dates from time to time. The character-building efforts of Arnold School and its Saturday night dances had not left Mike self-assured and with a sense of entitlement. Mike had the usual worries of a young man living on his own for the first time, but he had other anxieties as well. He abandoned religion (the university had no religious requirements) and Mike's apostasy troubled him from time to time. He was again in an unfamiliar educational institution that, despite the broadening of clientele over the previous few years, remained an exclusive institution where only one-third of the student body came from families of manual workers. Mike was one of the few underprivileged boys to have benefitted from reforms in the English educational system. He had new friends but the teaching staff recognized his rural accent and unpolished manners, and he felt this set him apart.

Mike had little choice but to make the best of his new situation. He was by now socially and emotionally far from the market gardens of Marton Moss, and when he visited home during the summer holidays he worked on local trains or at Blackpool's Pleasure Beach, one of the central amusement districts, rather than assist in his father's garden as might be expected. Several times he quit his jobs to travel, sometimes alone, hitching lifts from passing motorists to the Lake District or other destinations. Any lingering thoughts that Mike should spend his holidays working in the family garden were removed in 1953. Rowland had not been able to afford glasshouses for the new property he purchased in 1948, gambling that the weather would cooperate. After a few years of relative success, early frosts ruined the chrysanthemum crop and Rowland was forced to declare bankruptcy and give up his garden. He and Molly bought a house in urban Blackpool where Rowland took a job as night porter at the Queen's Hydro Hotel, eventually rising to Head Porter before retiring at age seventy-nine. Molly had already resumed her bookkeeping and worked with a garage until she was seventy-four. Rowland was not the last gardener to leave the Moss at a time when foreign competition was making market gardening increasingly unprofitable. For Mike, there was no returning to Marton Moss.[34]

Mike began a new stage of his life in 1953. In this year, he turned twenty-one and celebrated with a big party catered by his landlady, Mrs. Elton-Jones, and attended by family and friends from Blackpool and Manchester. Molly and Rowland came to celebrate with their son along with Mike's brother Robin who was now at Arnold School himself. (Robin later enrolled at the University of Manchester to study accounting.) Cousin Dennis, who continued in the market gardening business, also attended the party with his fiancée, Joyce, and two dozen of Mike's friends from university. For Mike, parties were important ways to reaffirm his friendships, especially if he could play the role of generous host. 1953 was also the year Mike graduated from the Honours Chemistry program. He was quite upset with his second-class standing, but it was a high second in a very demanding program that granted a first to only seven of the forty-three graduates. Holding back tears, he asked the department Head what to do next, fearing that he had ruined his chances for a career in chemistry. To his surprise, he was reassured of his ability in chemistry and told to continue with graduate work, proceeding directly to the doctoral program.[35] Fortunately, state grants were available for graduate students and Mike was able to continue with government support.

Mike joined a cohort of twelve doctoral students in the vibrant and exciting School of Organic Chemistry under the highly respected and well regarded Professor of Organic Chemistry, Ewart R.H. Jones. Graduate students participated in studies of vitamins, acetylenic compounds (including those produced by micro-organisms), terpenes (including compounds produced by fungi), steroids, and in various aspects of biochemistry. Nearby, students of physical or inorganic chemistry studied reaction kinetics, thermochemistry and bond dissociation energies, photochemistry and radiation chemistry, thermodynamics, and various theoretical topics. In all, over eighty research students assisted with projects in the wider department. The Labour government (1945–1951) had encouraged funding for basic, investigator-initiated research following the war while industrial chemical companies and the Rockefeller Foundation provided research grants to members of the Department of Chemistry. Under the supervision of H.B. Henbest, Mike studied steroids, a class of hormones that was attracting considerable attention at the time and would be used in such new pharmaceuticals as arthritis anti-inflammatories and the contraception pill. The Manchester group was one of a handful of laboratories worldwide racing to find efficient and inexpensive methods to synthesize the steroid hydrocortisone.[36]

The only formal requirement needed to complete the doctoral program was the completion of a research project and the defence of the written thesis, which seemed to Mike well within his abilities. But then Henbest, a brilliant but quiet young organic chemist, left on a year's leave of absence to Harvard University soon after giving Mike a research problem on an aspect of hydrocortisone synthesis. Professor Jones filled in as Mike's supervisor and provided stimulating and encouraging support. The doctoral students typically concentrated on their own work, but Mike took an interest in all the research going on around him. He demonstrated an extraordinary knowledge of everyone else's projects but, perhaps ironically, his own research did not progress smoothly. For a year all seemed to be going well, when quite unexpectedly, Jones read a recent publication on virtually the same problem that Mike was studying.

Suddenly, Mike's research was redundant and his past year's work had been wasted, made all the more distressing in his mind by the recent failure of a dating relationship. Mike became deeply depressed and lost interest in continuing his studies. In fact, he was so dejected that friends feared the worst.

Somehow he pulled through and, when Henbest returned to Manchester, began a new project to study the stability of the oxygen atoms on six-sided carbon molecules — cyclohexane diols. Yet circumstances were now very different. Not only did Mike have to complete a three-year project in two years, but Henbest was unable to provide him with the necessary emotional support. Henbest was a good scientist, but humourless and occasionally bad-tempered, especially with people he did not like. As a supervisor, he typically left students alone to do their work and showed an interest only when the project was going well. Mike was respectful, but the cool and distant relationship with Henbest did little to enhance his confidence as a researcher, and he continued to question his career prospects in chemistry. A particularly low point came in the autumn of 1955 when Mike was unable to carry out laboratory work for several weeks after breaking his foot in a nasty rock-climbing fall. Henbest arranged an interview just before Christmas to tell his student that his doctoral research was not going well, and that he might have to leave without his degree. This time, instead of being downcast as might have been expected, Mike responded with quiet anger and he vowed to finish.[37] Help came from Professor Jones who had recently been elected to the Waynflette Chair of Chemistry at Oxford University and had moved his laboratory and some of his staff with him. Professor Jones offered Mike some lab space, where he finished his research in the spring of 1956, away from Henbest.

During the fall of 1955 Mike made plans for the following year. Many of his peers went straight into industry — academic careers were still seen as the preserves of the socially more privileged — but others sought a further year of education working in a different laboratory. Large research institutions or funding councils often set aside a small amount of money for post-doctoral fellowships to provide a recent doctoral graduate an opportunity to study under the supervision of another established researcher. Other students had applied for fellowships; why not Mike? Like many of his friends, he hoped to obtain a position for a year in the United States before a career in industry. He sent out a number of applications before accepting a fellowship at the University of Colorado, but in May 1956 he discovered that the university had made a mistake and did not have the funds to support him. Mike had no other prospects and his hurried attempts to find one came to naught.[38]

Late in the spring of 1956, almost at the last minute, another student in the Department of Chemistry who had been studying biochemistry declined a post-doctoral fellowship at the British Columbia Research Council laboratories in Vancouver, Canada, to accept a more lucrative offer in the United States. The Vancouver fellowship was to study biologically important organo-phosphates, quite different chemistry from that Mike had been studying. Neither Vancouver nor the British Columbia Research Council were known as research centres and Mike had never heard of the supervising scientist. At least the fellowship sponsor, the National Research Council of Canada (NRC), had a good reputation. An NRC chemist had visited Manchester University a few years earlier, and one of Manchester's recent and promising chemistry graduates, John Polanyi, whose father had been a well regarded quantum chemist, had just finished two years as an NRC Post-doctoral Fellow.[39] Mike soon discovered that the supervising scientist in Vancouver, Har Gobind Khorana, was a rising young biochemist rapidly making a name for himself.

Despite any reservations he may have had, Mike wrote to Khorana about the fellowship and soon found himself at an interview in London with the Director of the Research Council, Gordon Shrum, who was touring Britain to recruit new academic talent. Shrum, also Head of Physics at the University of British Columbia, was a bold, determined administrator with a free hand to promote his projects and a man who expected great things of the people whom he hired.[40] Mike had only a few minutes to make a good impression, but he must have been successful, since, when the interview was over, Shrum recommended to Khorana that the young Englishman be given the fellowship.

Mike was going to Vancouver, but not before completing his thesis. The laboratory research was done, but it still had to be written up in an acceptable form. Because he was working long hours at the university and had spent his last few pounds to book passage across the Atlantic, he and his friend Ian (who was in a similar situation) decided to move into the Department of Chemistry library. The library had a splendid Victorian décor with big leather chairs perfect for sleeping, and the department secretary turned a blind eye to the squatters who stayed nearly two months. The two students also took jobs in a large commercial bakery to make some extra money, working two night shifts a week for several weeks until Ian fell asleep and an unsupervised sausage roll packaging machine ran amok. Both he and Mike were fired. Finally, in the summer of 1956, Mike completed his thesis, thanking Jones first for providing research facilities and dedicating it "to Mother and Father." He defended it at a public, oral examination to the satisfaction of two examiners, one from the University of Manchester and one from outside the university.[41] Although there had been tensions between Mike and Henbest, they soon faded since, after the successful defence, the two of them co-authored three publications.

Until this point, Mike's academic success had been guided by institutions designed to recognize and advance talent such as his. He had the benefit of a keen mind and a family that encouraged his education, considerable assets to be sure, and he did his part by working hard and doing well as a student. However, despite his educational achievement, Mike had not adopted a new social identity. He still preferred political debate and drinking beer in pubs — good working-class traditions — to drinking pink gins or whiskey and sodas in hotel lounges. He had acquired strong technical skills in chemistry but not the smooth, apolitical style of socializing so important to membership in the upper middle-class. He was not afraid to state his opinions bluntly amongst his friends, was wary of authority, and hated wearing a necktie, often refusing to do so, as evidenced by his graduating class photo. Mike's periodic lack of confidence and his emotional sensitivity also diminished his promise as a professional chemist and certainly as an academic. Henbest, for example, when appointed to a Chair at Belfast University, offered his student Ian a lectureship, but not Mike. Like the character Dixon in *Lucky Jim,* Mike was still out of place with the established academics.

Fortune now intervened to take Mike to a new world and new opportunities. During his hasty preparations for the long ocean voyage and train ride that would take him to Vancouver, Mike stopped in at the Canadian Consulate where an immigration officer gave him some advice. Instead of entering the country as a student, he should, the officer informed him, apply for an immigration visa in case he wanted to stay.[42] Mike took the advice, unaware of just how significant his proposed year in Canada would be.

2

A NEW LIFE

Michael Smith arrived at Vancouver's Canadian Pacific Railway station on September 23, 1956, with five dollars in his pocket. He stepped off the platform into a grey, overcast day, in a downtown that had changed little in the past twenty years. A few large, stylish hotels and office buildings suggested the city's wealth but many of the old commercial buildings of brick or concrete had a seedy appearance, and some sat vacant. The surrounding residential areas were not particularly attractive either, with many old, deteriorating wooden houses and a few new but unattractive highrises. Smoke from nearby sawmills did little to enhance the image but at least vehicle traffic flowed unimpeded on this Lord's Day. Low clouds obscured the mountains and sea to the north.[1]

This unflattering first impression of Vancouver reinforced Mike's intention to stay only a year before returning home to work, perhaps for an English pharmaceutical company. He also had his doubts about what the city could offer in the way of scientific experience. Would his laboratory have all the necessary physical and intellectual resources? Whatever Gordon Shrum had said, Vancouver was not a major research centre and it was far from prominent laboratories elsewhere. Mike was bold enough to travel to Vancouver, but he remained sensitive to how others treated him and he knew no one in this part of the world. Would he get on well with his new colleagues? Friendships were important to Mike, perhaps unusually so. Would his rural background and earthy sensibilities be a liability? He surely considered the matter of social class and its likely practical effects on him. A year of post-doctoral work that provided a good technical education and satisfied his personal needs could transform Mike's ambitions forever.

As a post-doctoral fellow at the British Columbia Research Council, Mike was fortunate in being able to spend his time in research with no administrative or teaching commitments. This meant that a well-equipped laboratory directed by an established scientist was of crucial importance. Mike had good reason to be apprehensive about the scientific standing of his destination because although the Canadian government supported small research stations in agriculture, forestry, and fisheries, and Vancouver was home to the provincial university, the local research community was poorly developed and could provide little advanced education for aspiring scientists. The University of British Columbia had pockets of active researchers but it was only beginning to develop the infrastructure necessary for advanced research and graduate studies. Mike, however, was not even heading to a UBC laboratory but rather to one at the little-known British Columbia Research Council, a highly unlikely institution to provide a first-rate post-doctoral experience.

The Research Council was never intended as a place of basic research or post-doctoral education. It was formed as an applied-science arm of the university during the Second World War to solve problems for industrial clients. Research Council investigators examined, for example, steel tempering, the feasibility of extracting gas fuel from sawdust, feeding standards for fur-bearing animals, irrigation methods, rodent repellents, industrial efficiency, and the treatment of mastitis in dairy cattle. In the early 1950s, the most celebrated research projects were those to control metal corrosion and marine borers, a small, wood-boring bivalve that ravaged the docks, boats, and log booms of coastal industries. Political debate over the Research Council centred on whether the modest government grant was properly spent to support practical and economically valuable research.[2] At least the B.C. Research Council was on the University of British Columbia campus and in fact moved into new buildings in 1952 which later became the university's Geological Sciences Building. This kept the Research Council close to scientific developments at the university.[3]

However unlikely the B.C. Research Council was as Mike's postdoctoral destination, his supervisor was an extraordinary and powerful mitigating factor. Laboratories are people as well as places, and Har Gobind Khorana was not simply a rising young scientist but a commanding intellectual force in the emerging field of molecular biology, a specialized facet of biochemistry that paid particular attention to the role of nucleic acids. Mike may have credited chance for his own journey to Vancouver, but Khorana's appearance in the city was just as fortuitous. Four years before Mike's interview in London, Shrum had met Khorana in England, brought him to Vancouver, and provided him with unqualified support.

Both Mike and Khorana were lured to Vancouver during the first years of a great transformation at the University of British Columbia. Gordon Shrum played a prominent role in the movement to make the university into a great centre of research and graduate education, extending its earlier work as an undergraduate, teaching institution. To some, Shrum was a superb manager and charismatic teacher. To others he was "a big, bold man with a booming voice," ambitious, occasionally ruthless, quick, and impatient — an intimidating and sometimes frightening person on a mission to promote science and scientific research at UBC.[4] Shrum had broad administrative influence at the university and in the province's and nation's scientific community. In 1951 he accepted the position of Director of the B.C. Research Council and, because the National Research Council had recently built the Atlantic and Prairie Regional Laboratories but nothing on the west coast, wangled funds from the NRC for a special project at the B.C. Research Council labs. Federal research funding in the early Cold War era was growing as part of a new national commitment to science, encouraged by NRC president Edgar Steacie who was a great supporter of investigator-directed basic research to advance knowledge. He provided a boastful Shrum with a modest $12,000 (less than a top UBC salary) to launch a new laboratory.

Shrum had a challenge to meet, and his success would provide the opportunity for Mike's fellowship. In the summer of 1952, while visiting Great Britain, Shrum heard from Professor A.R. Todd at Cambridge University of a brilliant young man named Har Gobind Khorana who might be interested in relocating to Vancouver. Like Mike, Khorana was from a poor family and had worked hard to overcome social and economic obstacles to become a scientist. Born in India, Khorana first studied chemistry at the University of the Punjab and then earned scholarships for graduate degrees in England. Following a post-doctoral year in Zurich, Switzerland, financed from his own careful savings, Khorana returned to India to find that local politics had made employment at his home university unlikely. He returned to England for a three-year fellowship at Cambridge University where he established himself as a bright and very promising scientist, but his family background made academic promotion in England almost impossible. Shrum hesitated but then realized what an opportunity he had; Khorana was clearly an exceptional talent who was willing to work in Vancouver for low pay. In lieu of a good salary and well-equipped facilities, Shrum promised Khorana "all the freedom in the world."[5]

Khorana immediately married his Swiss sweetheart, Esther, and travelled to Vancouver to establish his laboratory. During the four years before Mike arrived, Khorana impressed Shrum with his soft-spoken honesty, sincerity, and utter commitment to science, "working morning, noon, and night — and in a hut." (Khorana later moved his laboratory into the west wing of the new Research Council building.) Khorana soon earned international praise for developing a new branch of chemistry to synthesize molecules of biological importance.[6] Members of Khorana's lab devised a new method to synthesize ATP, a compound that stores energy in living organisms. In 1954, his lab synthesized uridine monophosphate, a key factor in cellular growth and important in studying enzymes. Local newspapers hailed the new procedure as a significant step toward a cure for cancer, forcing an embarrassed Khorana into hiding for a few days until the excitement passed. His central interest, however, was the synthesis of oligonucleotides, short strings of nucleotides that are the building-blocks of nucleic acids, including deoxyribonucleic acid (DNA) which is the genetic blueprint of an organism. While Khorana did his impressive work, Shrum lobbied the provincial government for greater Research Council funding.[7]

Having studied organic chemistry rather than biochemistry or molecular biology, Mike knew little about Khorana or the stature of his laboratory. Mike was unaware that two rising American biochemists and future Nobel laureates, Arthur Kornberg (who later purified and characterized an enzyme capable of synthesizing DNA) and Paul Berg (who developed recombinant DNA technology), were visiting Khorana in the summer of 1956 to study his new techniques of synthesizing nucleotides. Protein and nucleic acid research was making great advances during the "revolution in biology" that applied the tools of physics and chemistry to living organisms. DNA was now known to be the genetic material of a cell, and Watson and Crick's recently proposed double-helix, ladder-like structure of the molecule inspired Khorana and other biochemists to investigate how DNA (or, more precisely, messenger RNA, which is a copy of DNA) encoded the proteins that constituted an organism.[8] Khorana's work was as good as could be found anywhere.

Mike's entrance in Khorana's lab, however, was disappointing. He quickly realized the complexity of Khorana's work and how much he would have to learn. Not knowing anyone in the lab, he began to doubt his abilities and the wisdom of accepting a fellowship in this remote location. Khorana soon assigned him a research problem related to oligonucleotide synthesis but when left to work out the experimental details Mike floundered. For several months none of his experiments worked properly. He also panicked when, soon after his arrival, he was asked to address the local

Biochemistry Discussion Group, an informal collection of scientists from the Research Council, UBC, and nearby government laboratories. He paced back and forth nervously while waiting for his turn to speak, and when the time came he was awkward and tongue-tied. Not surprisingly, Khorana was unimpressed with his new research fellow. The problem might have been Khorana himself, whose vast intellect could intimidate as well as inspire. However, Mike was certainly not defeated. Just as he had persisted at Arnold School and the University of Manchester to succeed in unfamiliar environments, he set about learning the new chemistry and by the end of the year he was designing successful experiments that extended Khorana's work. As he began to achieve good results, Mike's confidence and enthusiasm rose.

In fact, Mike soon realized that he was an integral member of a dynamic team at the forefront of scientific research. Khorana made local headlines and international news early in 1957 with the announcement that his laboratory had won an $80,000 grant from the United States Public Health Service to investigate the chemical substances in living cells. The grant was huge by UBC's standards (it was $30,000 more than the research program directed by Marvin Darrach, the Head of UBC's Department of Biochemistry), and the U.S. agency did not often provide awards to foreign laboratories.[9] The funding assured members of Khorana's group that they could participate in leading-edge research to extend knowledge in their chosen field.

Although it took several months, Mike found Khorana to be an impressive mentor and "a quiet, unassuming yet compelling scholar who inspires ambition and confidence." The young post-doctoral fellow marvelled at his supervisor's capacity to remember details about all the work in the lab and provide informed guidance. They both understood what it meant to grow up in a poor, rural family and the difficulty of advancing a scientific career in the face of prejudice. Once Mike began to prove himself in the laboratory, he spent many hours with his mentor talking about science and the synthesis of oligonucleotides, procedures that would soon help to explain the mechanism by which the genetic template of DNA directed the formation of proteins and hence the development of an organism — the as yet unexplained genetic code. Mike discovered a new passion for his work along with a deep respect for Khorana, and for the first time he considered a career in academic research.[10]

Much to Mike's relief, he found other members of the research team to be just as congenial. One of them, Gordon Tener, was born in Vancouver and educated in the province's schools. Mike realized that British Columbia had talent of its own and a public (state) school system capable of developing it. He and Tener became close, lifelong friends and colleagues, sharing ideas, supporting and encouraging each other, and occasionally visiting the Vancouver beer parlours together. Mike also formed friendships with other young scientists at the University of British Columbia, sharing a basement suite with other post-doctoral fellows in UBC's Department of Chemistry. These young chemists (many of them "Brits" from England or the Commonwealth) played sports — Arnold School had left Mike fond of grass hockey, although he was still not particularly talented — and held wild Saturday night parties with plenty of drinking. Khorana was initially a little dismayed with these new activities but could do nothing to dissuade his sociable protégé and sometimes joined him for a beer.

With new confidence, Mike reintroduced himself to the Biochemistry Discussion Group. The discomfort and awkwardness that afflicted him at his first formal address quickly disappeared as he became an active and dedicated participant. The Discussion Group usually met at the home of Jack Campbell, a UBC professor of dairy microbiology, where Mike visited with friends long after the formal program had finished. These new colleagues provided all the social acceptance and friendships that Mike needed.

Not only did Khorana's laboratory prove to be a stimulating place for science, Vancouver soon turned out to be an ideal city for someone like Mike. The clouds that greeted his arrival eventually parted to reveal a magnificent natural setting that easily compensated for the drab city. Like Blackpool, Vancouver was a temperate and humid coastal city with a population at this time of about

350,000. Unlike any place in England, Vancouver sat against a splendid backdrop of mountains that fuelled Mike's excitement for hiking, climbing, and camping. The city's strong British roots and large population of expatriates provided familiar traditions and a recognizably Anglo-Canadian lifestyle. Vancouver had cricket and rowing clubs, fish-and-chip shops, British literature, tweed jackets, social clubs, large Anglican and Presbyterian Churches, and plenty of British accents. Vancouver flew the Union Jack with the rest of Canada, and the city had just hosted the exciting 1954 British Empire Games. The city had a Stanley Park, as did Blackpool, an English Bay, and an exclusive neighbourhood in nearby West Vancouver called the British Properties developed two decades earlier by the Guinness family of Irish brewing fame. Vancouver's cultural amenities were modest but growing: the city had a symphony, theatre, art gallery, and museum.[11] Although Vancouver lacked English-style pubs, liquor regulations were easing after decades of tight control. Licensed hotel beer parlours still segregated "gentlemen" from "ladies and escorts," but after 1954, cabarets and lounges could serve drinks *and* provide entertainment. Local breweries prospered.[12] Mike loved hiking in the mountains and could enjoy a symphony concert or share a glass of beer with friends.

Vancouver was also receptive to energetic and ambitious young men who were eager to become involved in new developments. Mike was less restricted than he would have been in Britain to pursue the social and economic mobility promised by thirteen years of advanced education. Behind its veneer of civility Vancouver was a brash young city that had grown quickly after incorporation in 1886 and the arrival of the nation's first trans-continental railway. As the industrial and financial centre of the province and western Canada's leading commercial city, Vancouver had long enjoyed a free-enterprise business ethic that encouraged real estate deals, stock speculation, colourful entrepreneurs, and nouveaux-riches additions to the city's elite. Vancouver's Non-Partisan Association, supported by well-established commercial interests, had dominated civic politics for decades and was making plans to reinvigorate the city. A Social Credit government had a firm grip on provincial politics and was pursuing its vigorous if unorthodox agenda of development. Many immigrants who came to Vancouver in search of work and better living conditions found them during the reassuringly long post-war economic expansion that faltered only infrequently, as in 1958, 1973, and 1983. Vancouver was for many newcomers a place where hard work and intelligence paid off, regardless of family background, although there certainly were protected circles of power and affluence. Like the city and province, the scientific community, centred on the provincial university, was growing and transforming in ways that would reinforce Mike's interests and outlook.[13]

Of course, Vancouver had an unattractive side, but nothing beyond what Mike had already experienced. The city retained a frontier roughness. Sawmill smoke and polluted beaches sometimes tainted the natural beauty and Vancouver crime statistics were generally above the national urban average, with particular problems in drug trafficking, prostitution, drinking, and gambling. Riotous behaviour occasionally tarnished sports meets and other festivities. Vancouver was home to longstanding and widespread prejudice, particularly against immigrants from Asia or southern Europe and their descendants. Native people lived in poverty at the margins of Vancouver society and women were expected to fill traditional social roles. But as elsewhere, post-war social attitudes in Vancouver were changing. Labour unions, middle-class reformers, and, since the 1930s, the social-democratic Cooperative Commonwealth Federation (that became the New Democratic Party in 1961) had been left-leaning influences in civic politics that opposed certain features of free-enterprise. Certainly the ills of Vancouver were no worse than those of glitzy Blackpool or industrial Manchester, and Mike was not alone with his socialist sympathies.[14]

The immigration officer who had provided advice in London had made a useful suggestion to Mike, who, within a few months, had decided that he very much wanted to stay in Vancouver for more than a year. He had exciting work and new friends in an enjoyable part of the world, just about everything he needed to feel good about his life. Post-doctoral fellows in Khorana's lab came and went, but Mike remained for four years to enjoy all that his new home could offer. He learned to ski — he loved skiing and worked hard to develop his skill — adopted local slang (such as "golly" and "gee"), and began reading the *New Yorker.* He took several travel holidays with his friends and four of them once went to Mexico City, driving as far as San Diego and camping along the way before flying to their final destination. Mike demonstrated his generous spirit by cooking many of the campground meals for his friends, who were impressed with his culinary skill. Having always wanted a sports car, Mike finally bought a used, white Triumph TR3 with red upholstery that he pampered and often drove at illegal speeds. He made numerous friends who enjoyed his fun-loving, good natured disposition and overlooked his rough edges.

Mike was particularly pleased to be dating Helen Wood Christie, a student of dairy microbiology who worked in Jack Campbell's laboratory next to the Research Council. Helen was the daughter of a former UBC professor and Head of Forestry, the sort of person the son of a market gardener would be very unlikely to have known back home. They first met while out with separate friends in the classy beer parlour of the Georgia Hotel in downtown Vancouver, although Mike might have seen her playing touch-football games at UBC. One of Mike's friends must have known one of the women, permitting both groups to meet in the area of the hotel's beer parlour reserved for ladies and escorts. Mike was nervous at their first meeting, compulsively peeling the label from a beer bottle as he chatted with Helen, who found him pleasant and amusing. They shared interests in music — their first date was to hear the famed cellist Pablo Casals — and the outdoors. Helen liked exploring the mountains as much as Mike did, and she was an excellent, graceful skier. Over the years, their relationship grew more serious.

Mike enjoyed life in Vancouver immensely. He put considerable energy into his social and recreational activities, but he put even more into his work with Khorana. Mike studied the fields of biochemistry and molecular biology to learn what was already known and what problems needed solving next. He familiarized himself with simple but effective laboratory procedures. For example, he learned how to extract biologically important compounds from plant or animal tissue, although the new techniques of chemical synthesis devised in Khorana's laboratory slowly replaced the need for natural specimens. The lab used homogenizers as simple as a mortar-and-pestle and tissue grinders that resembled kitchen blenders to break cells apart physically so that the internal parts could be fractionated (separated using centrifuges) and analyzed.

Mike also learned to use chemical solvents to separate biological compounds. Different molecules would separate from each other by relocating (partitioning) to one particular solvent over another. In one common procedure, Mike placed the initial compound and a solvent at the top of a glass column that was packed with an absorbent material. He poured additional solvent through the column, causing the compound to seep through the packing material for collection as it dripped out the bottom. Because different molecules partition differently between the solvent and the packing material, they appear at distinct intervals at the bottom of the column. By collecting the outwash in a series of fractions, Mike collected the separated molecules. Soon after, Khorana's laboratory invested in an automatic collecting device that was essentially a wheel that rotated test tubes under the dripping column. This procedure, column chromatography, was used extensively by members of the lab.

In addition, Mike learned to use "high tech" tools such as the Cary spectrophotometer. The spectrophotometer contained a prism that sent a narrow wavelength of light to a photocell through a sample held in a quartz cell. By measuring the percentage of light absorbed by the sample, Mike could determine the properties of the sample and quantitate it. (For example, nucleotides absorb ultraviolet light because they have multiple double bonds. When a sample is placed in a beam of UV light, the amount of absorption indicates the concentration of the nucleotide.) Electrophoresis was another common procedure that Mike used to separate compounds. In this technique, a mixture of compounds is placed on one end of a strip of paper saturated with a dilute salt solution, and an electric current run through the strip. Because charged molecules are pulled along the paper strip at different rates, the compounds separate from each other.

Needless to say, with the use of animal tissues, solvents, and other chemicals, the B.C. Research Council biochemistry laboratory was a smelly, noxious, and occasionally dangerous place. Once, while Mike was working on a group of chemicals whose function was to control chemical reactions (the methoxyl-trityl family of protecting groups for nucleoside-5'-hydroxyl groups), a synthesis of trimethoxytritanol erupted. Mike was found the following day scraping the residue from the ceiling onto a glass plate and contemplating the large orange stain that remained. Such accidents compromised the experimental results but he now knew how to learn from his mistakes. Mike was more amused than embarrassed, and despite the accident he went on to devise a useful procedure that is still used for automated synthesis of DNA and RNA fragments. This and other successes in the laboratory buoyed Mike's spirits and fired his enthusiasm.

One of Mike's most significant scientific contributions was to help develop the phosphodiester method of synthesizing oligonucleotides. Beginning with purified nucleotides, the first step was to initiate a series of chemical reactions to attach "protective" molecules to prevent further, unwanted side reactions. Subsequent reactions joined the nucleotides together repeatedly to build the oligonucleotide with a desired sequence. In between each reaction, the synthesized oligonucleotides were purified by washing in a column with a saline solution. The protecting and coupling reactions were not very efficient, requiring a large initial quantity of nucleotide reactants. In addition, Mike helped develop a general procedure for the preparation of nucleoside triphosphates, including ATP, and nucleoside-3',5' cyclic phosphates, a class of compounds whose existence and great biological significance in intracellular regulation of metabolic processes was just becoming understood.

For years the B.C. Research Council under Shrum's directorship provided a proud home for Khorana's internationally recognized research and Mike's glory days as a post-doctoral fellow.

Shrum was quick to point out the importance not only of this particular research, but of basic research more generally as the vital foundation of applied science.[15] Khorana and his doctoral student in the UBC Department of Chemistry, John Moffatt, caught the world's eye in 1959 with the synthesis of co-enzyme A, a molecule involved in the production of energy for metabolism. This breakthrough was hailed as "a beautiful piece of chemical artistry." Instead of acquiring co-enzyme A from yeast at $17,000 an ounce, it was now available synthetically at a fraction of the cost. The Council's annual report also noted in 1959 that Smith was among those who had found a procedure to combine nucleotides into longer chains of polynucleotides, and in 1960 that Smith had recently synthesized a nucleoside cyclic phosphate of the type known to participate in generating muscular energy.[16]

Yet the halcyon days at the B.C. Research Council were numbered. Shrum knew that Council facilities were limited and that Khorana's reputation was now well established across the continent. The star biochemist might leave and take his laboratory personnel with him. Rumour had it that Shrum tried to find Khorana a permanent appointment at UBC, but the Departments of Chemistry and Biochemistry (where Khorana had been a part-time lecturer in the spring of 1956) were uncooperative.[17] Instead, Shrum used his powers as the new Dean of Graduate Studies to have Khorana appointed to a novel position as University Professor, without term, and at a top salary of $13,000 per annum. His salary initially came from a Graduate Studies contingency fund with hopes for subsequent funding from the NRC, not the university.[18] The university had recently raised $11 million (with matching provincial money) for new buildings, but no one managed to find funds to support Khorana's expensive research program.

In an era when UBC's administration still could — and did — make autocratic staffing decisions, it seems odd that UBC was unable or unwilling to find a more secure place for this obvious asset.[19] As non-scientists, President Norman MacKenzie and other administrators may not have fully appreciated the talent before them. Friends of Khorana believed, however, that ethnic prejudice

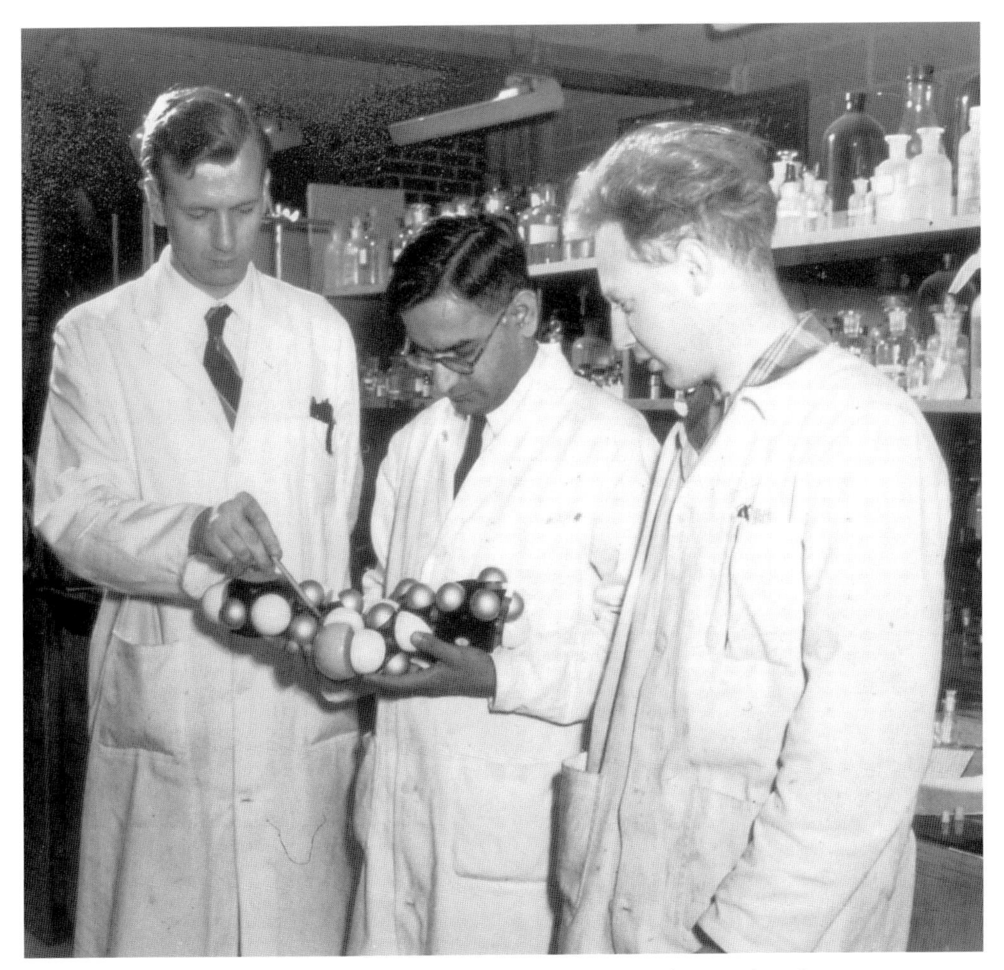

▲ Mike (right) and Peter Gilham (left) consult a molecular model held by Har Gobind Khorana in the B.C. Research Council Laboratory. (COURTESY UBC ARCHIVES)

▲ Khorana's B.C. Research Council Lab with experiments continuing late into the night. (COURTESY UBC ARCHIVES)

▲ Paul Potter and Mike in their basement suite sampling local brew. (COURTESY HELEN SMITH)

▲ Mike washing his beloved TR3 sports car outside the B.C. Research Council building. (COURTESY HELEN SMITH)

▲ Paul Potter (left), Dave Frost, and "Smithy" on a road trip to Mexico, c. 1958. (COURTESY HELEN SMITH)

▲ The Biochemistry Discussion Group, 1960. Kneeling, from left, are John Vizsolyi, Bill Razzell, Gordon Tener, and John Moffat. Seated are Dave Idler, George Drummond, Sid Zbarsky, Eli Reichmann, Har Gobind Khorana, Jack Campbell, and Hugh Tarr. Standing (first row) are George Strasdine, Gunter Weismann, Michael Smith, Ray Ralph, Margaret Duncan, Audrey Ells, Nora Neilson, Jim Polglase, Al Paterson, and Henry Tsuyuki. Standing (second row) are Tony Warren, Dave Rammler, Phil Townsley, Berkhard Lerch, Ian Caldwell, Neil Tomlinson, and Bob MacLeod. (COURTESY GORDON TENER)

▲ Mike and Helen on their honeymoon visit to England, with Mike's parents Molly and Rowland Smith in Blackpool, 1960.
(COURTESY HELEN SMITH)

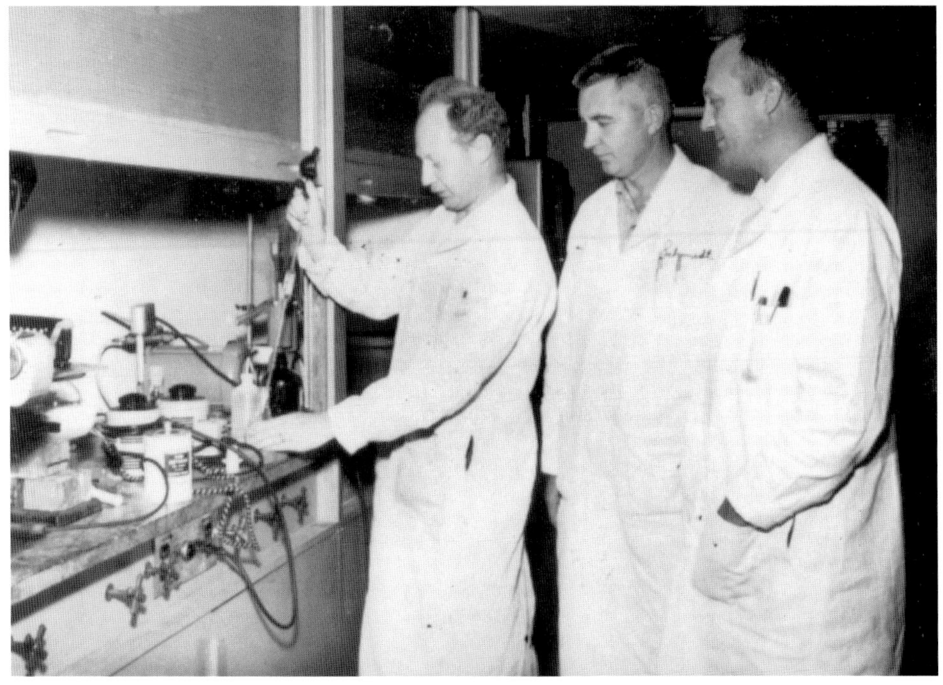

▲ Mike demonstrating experimental techniques at the Fisheries Research Board in Vancouver. (COURTESY PETER SCHMIDT)

▲ Mike holding a research crab while visiting Irving Goldberg at his lab in Chicago. (COURTESY UBC ARCHIVES)

▲ Mike skiing at the Mt. Baker ski area, 1960s.
(COURTESY HELEN SMITH)

▲ Mike with Helen and their children Tom, Wendy and Ian at home for Christmas, c. 1965. (COURTESY HELEN SMITH)

▲ Mike with one of his smaller catches.
(COURTESY HELEN SMITH)

played a role, and the scientist who had been denied a career in India and Britain again felt "out of place everywhere and at home nowhere." Except for one research associate in the Department of Physics, no faculty members in Medicine or Science — perhaps the entire university — were of Indian descent in 1960. Vancouver, like British Columbia and Canada more generally, had a strong history of discrimination that was only just beginning to change.[20]

Shrum also knew that most of the funding to support the internationally renowned research team that included Dr. Michael Smith came from the United States. (The exception was Gordon Tener whose salary came from the Research Council.) To encourage Khorana and his team to stay at UBC, Shrum and local philanthropist Walter Koerner formed a committee of business leaders and university personnel to identify possible funds for a "Khorana Scientific Institute," with the federal government considered as a source for nearly half of the required $500,000. Shrum reminded those who would listen that Canadian funding had begun this important work, and Canadian funding should continue it as a symbol of Commonwealth cooperation among India, England, and Canada. He further prophesied — correctly — that Khorana was well on his way to winning a Nobel Prize, a tribute that belonged to a British Columbia laboratory.[21]

Yet the committee was not charged with raising any money and it created no new institute. When a year later the University of Wisconsin offered Khorana stable funding, a university position, and excellent laboratory facilities with a large staff in the Institute for Enzyme Research, he accepted. He felt compelled to work in the country that provided him with his research money. Besides, colleagues thought that Khorana was looking for a change in order to feed his insatiable appetite for knowledge. Several of Khorana's committed and loyal lab members prepared to move to Wisconsin with him, including Mike who had no other immediate job prospects. No doubt with great understatement, Shrum noted in the Seventeenth Annual Research Council Report that "the decision of Dr. H.G. Khorana to move with his group of organic chemists to the University of Wisconsin was universally regretted."

Before Mike left for the United States he celebrated an important occasion. He and Helen were married on August 6, 1960, on the front lawn of her parents' home near the town of Mud Bay on Vancouver Island. Mike had suggested a civil ceremony by a Justice of the Peace located in an old building near the Vancouver General Hospital, but as this was next door to the Venereal Diseases Clinic Helen's mother insisted that their home on Vancouver Island was a more appropriate location. Mike agreed, providing the ceremony was not overly religious. Rev. Peter Kelly, a United Church Minister from the nearby town of Parksville who was from the Haida nation, conducted the wedding amidst plenty of flowers and sunshine. Gordon Tener served as the best man while Helen's two sisters comprised the bridal party. More than eighty guests attended and six of Mike's friends composed and sang a song describing some of "the groom's lesser qualities." They also suspended his car on blocks of wood, leaving him and Helen to lower the car before departing for the mountains of the Olympic Peninsula and then to Garibaldi Provincial Park. Following the camping honeymoon, Mike and Helen drove across Canada, sailed to England to meet family who had been unable to attend the wedding, and returned to North America to settle in Madison, Wisconsin.[22]

Mike's stay in Madison was short. At the outset, he was promoted to a position of research associate to continue working on the chemical synthesis of oligonucleotides, but this new role was no longer enough. As a husband, Mike had new financial responsibilities that could not easily be covered by his low salary, and before long he and Helen were expecting their first child. As a young scientist who had spent four years as a post-doctoral research fellow, he was ready to direct his own laboratory and use to the fullest extent possible the considerable knowledge and skill he had learned from Khorana. He had published five papers with his mentor and had met a number of prominent nucleic acid scientists who now recognized him as a promising young researcher in the small community of North American biochemists. He appeared not to have any grand ambitions, but he liked his independence and wanted to set his own research agenda, participating as an equal in the scientific community. Nothing would be better than to return to Vancouver where he and Helen had friends, mountains, and the ocean, if only Mike could find a suitable job there.

His prospects in Vancouver were not very good, and a few of Mike's friends suggested he look elsewhere for employment. It was still not an important centre of biochemical research, and prominent academics elsewhere were showing interest in Mike. Arthur Kornberg, for example, was believed to have invited him to Stanford University but the message never made it through. Mike thought he might apply to foreign universities but, believing that there was little interest in his particular expertise, decided that he would work at any research institution that wanted his skills. British Columbia had several laboratories outside the university that might have a place for him, any of which would be fine. Perhaps the prospects of leaving Khorana's lab where he had been so long had stirred his dormant insecurities, encouraging him to find any opportunity to return to where he had felt so comfortable.

After a few months in Madison, Mike heard from scientific colleagues in Vancouver that a research position in a local government laboratory might be available, and by January of 1961 he was in touch with the Vancouver Technological Station of the Fisheries Research Board of Canada. Dave Idler, who was known as an active member of the Vancouver Biochemistry Discussion Group, was the Head of the Chemistry Division who had recently been promoted to Director of the Halifax Station. He thought Mike would make an excellent successor. The Director of the Vancouver Station, Hugh Tarr, agreed and encouraged Mike to return as a Senior Scientist and Head of the Chemistry Division, a large and productive unit, for what promised to be enjoyable and creative work. Tarr suggested an annual salary near the upper end of the government scale — $9,100 — and promised to expedite the necessary paper work. Mike deeply appreciated Tarr's offer and warmly thanked him, wasting little time in submitting an application. Khorana provided a reference, stating that his former student had a very good background in chemistry and was intelligent and sound in his thinking. Besides, Khorana added, Mike was extremely pleasant and cooperative. The only problem, Mike half-joked, was that he might have to hitchhike back to Vancouver since expenses from the marriage, trip to England, and set-up of a home in Madison had left him broke. In March, Mike tendered his resignation to Khorana, leaving, he said, "with more than a little sadness."[23]

The Fisheries Research Board was a venerable Canadian institution concerned with the commercial productivity and conservation of the nation's fisheries. Established in 1898 as the Board of Management of the Marine Biological Station, it could claim to be among the country's oldest national research organizations, predating many of the country's universities. Research at the FRB began almost immediately with a temporary (and transportable) research station established at St. Andrews, New Brunswick, that was soon replaced by a permanent station. The Board also established laboratories on Georgian Bay in Ontario (1901) and at Departure Bay near Nanaimo on Vancouver Island (1908). The technological research station established at Prince Rupert in 1924 moved to downtown Vancouver in 1942 where it occupied an old soft drink bottling plant before moving to the UBC campus in 1959. There it sat on a bluff near the ocean and a five-minute walk from the B.C. Research Council laboratories.[24]

Mike knew that the FRB had a mandate to conduct applied research relevant to the west coast fishery but he also knew that this agenda was expanding. In its early years, FRB scientists were mostly visitors from universities who investigated problems to advance knowledge in their respective fields. This changed as the FRB hired its own staff and emphasized research of commercial significance, especially at the technological stations. By the time Mike joined the Vancouver Station, the FRB was again encouraging basic research and liaison with universities, and the technological stations broadened their research into physiological, biochemical, and bacteriological studies. The FRB also hosted visiting scientists working for the International North Pacific Fisheries Commission and received federal funding as part of the enhanced national research agenda that accompanied Cold War anxieties. (FRB administrators complained, however, that national funding was more generous to physics and engineering than food and biological sciences.) Mike had a few administrative duties to oversee a small staff of junior scientists and assistants (only one other member of the Division had a PhD) and to ensure the smooth operation of his laboratory, but he was largely left alone to conduct biochemical research.[25]

The FRB had traditionally restricted its mandate to problems concerning the ocean environment and marine biology, and Mike was happy to continue the work left by his predecessor. Dave Idler had been researching the energy expenditures of sockeye salmon, a major commercial fish, during runs up the Fraser River to spawning beds. Mike continued with a study on the feeding habits and survival of spawning salmon. He resumed Idler's other research to identify the odours that guided salmon to their birth streams (to help determine how commercial bait could be more aromatically attractive) and to isolate and purify the pituitary hormones of salmon to determine their physiological effects.[26] Mike had never really studied marine biology, but this proved to be no deterrent as he continued with these inquiries. He was not afraid to get his hands dirty, capturing his own marine animals and removing various organs. Although his dexterity was not the best, he became adept at using a hollow drill to extract pea-sized pituitary glands from salmon. Mike enjoyed field trips to beaches, seashores, and hatcheries from British Columbia to Oregon State to obtain lab specimens from salmon, crabs, octopus, oysters, or other marine creatures. With supplies from the field, he returned to "the bench" for analyses that resulted in scientific papers.

Mike was not content, however, simply to conduct marine research already begun at the Vancouver Technological Station. His interest in DNA and molecular biology in general was unquenchable and he returned to Vancouver determined to conduct research on nucleic acid synthesis, a hot topic of study as scientists on both sides of the Atlantic elucidated the genetic code. Less than a month after returning to Vancouver, Mike applied successfully for a United States Public Health Service Research Grant to pay for a post-doctoral fellow, E. Reiner, to work on the chemical synthesis of oligonucleotides and the structure of natural deoxynucleic acids. Such studies, Mike assured his administrative supervisor, would investigate fundamental problems in science and be relevant to the FRB's interest in energy utilization and the enzymatic synthesis of nucleic acids in fish tissue.[27]

Mike evidently had learned to write effective grant proposals, a necessary skill in the operation of a laboratory. He could communicate the rationale for and uniqueness of the proposed research, the basis of the research in existing knowledge, and the appropriateness of the research to the mandate of the funding agency. He knew when to make bold claims and when to be modest, how to demonstrate his capacity as a researcher, and how far he could challenge existing orthodoxy. This valuable writing skill enabled Mike to continue receiving grants for the rest of his career and publish the results of his experiments.[28]

Of course, Mike returned to his informal collegial network, and by the fall of 1961 he was with his old friends in the Biochemistry Discussion Group. He helped organize presentations for faculty members and graduate students in the UBC Departments of Biochemistry, Dairying, and Zoology, and scientists from the Cancer Institute of B.C., the Federal Department of Agricultural Science Services, and various laboratories on Vancouver Island. Mike still thought that Vancouver was isolated from the main scientific centres and hoped that distant scholars could be encouraged to visit. He asked Tarr if the FRB might sponsor two or three visiting speakers each year.[29] Mike may have been correct about Vancouver's distance from other research centres, but the city did host prominent intellectuals from time to time. The Nobel Prize-winning chemist Linus Pauling, for example, stopped in Vancouver soon after receiving his 1962 Nobel Peace Prize and chatted briefly with Mike, who was very excited.[30] Whatever limitations he felt there were in Vancouver, Mike was loyal to his new home and became a Canadian citizen in 1963.

Khorana also remained in Mike's informal circle of scientific colleagues. Perhaps more accurately, Mike was included in Khorana's growing network of former but devoted post-doctoral fellows. The two regularly corresponded, sometimes as friends (Khorana teased his former student that beer at the Madison seminars was strictly rationed) and sometimes as respected associates. Khorana sent Mike research papers for scrutiny and was pleased to receive critical comments in return. By 1966, the genetic code was largely explained and Khorana was recognized for his contributions by sharing the 1968 Nobel Prize in Medicine or Physiology with Robert Holley and Marshall Nirenberg. Over the years, Mike and Khorana met at conferences, visited in Madison when the opportunity arose, and sent each other visitors. Khorana was able to visit Vancouver in 1965 to talk to the Biochemical Discussion Group, and returned on many subsequent occasions.[31]

By 1963 Mike needed assistants from the University of British Columbia to help him conduct research for which he had funding. His interest in graduate students was not simply to find workers for his lab and perhaps to recreate the social atmosphere he had enjoyed a few years earlier, but also to initiate younger scientists into the whole scientific enterprise. Mike's lab was more diverse in the kinds of students it attracted than the culturally homogenous labs he had known in England and at the B.C. Research Council. One of his graduate students, Vivian Wylie, was from Tobago, West Indies and of African descent. Another of the FRB post-doctoral fellows in the laboratory was Japanese (who, many years later, had his grandson named in Mike's honour).[32] UBC and the province more generally were in a new era of cultural diversification and Mike increasingly worked with people from varied backgrounds as he looked for talent regardless of cultural origin to enlist the best laboratory help he could find.

His other student, Nadine Wilson, was one of the few women working on a UBC graduate degree in science. She had an undergraduate zoology degree from UBC and had subsequently worked as a laboratory technician for the Red Cross and at the university. Soon after she returned to UBC for doctoral studies her supervisor changed universities and left her looking for a new advisor. Wilson heard about Michael Smith from a young professor named David Suzuki (whom Mike described to Khorana as "a *Drosophila* geneticist in Zoology . . . who is modern and excited about things") and approached the Fisheries scientist for guidance.[33] Mike had worked with women as assistants, technicians, or support staff, but senior scientific colleagues in England and in Canada for the most part had been male. At UBC and universities more generally, women were poorly represented in many fields, especially engineering and natural sciences like physics or chemistry. The biological or life sciences at UBC, including biochemistry, were somewhat more friendly to women, both as students and as faculty members, perhaps because of their perceived relevance to such traditionally "female" fields as nursing and home economics.[34] (See Table 1.) Mike, perhaps thinking fondly of his mother, was impressed with hard-working, intelligent women and he was happy to have Wilson as a student studying salmon hormones.

TABLE 1
Female Science Graduates, 1963, 1965

1963 (Total Science Graduates: 197)	Female graduates
Life Sciences	10
Natural Sciences	4
Both	1
No Major	9
Total	24
1965 (Total Science Graduates: 296)	**Female graduates**
Life Sciences	21
Natural Sciences	9
Both	2
No Major	9
Total	24

Source: *UBC Totem*, 1963, 1965.

Wilson was the first woman to work toward a graduate degree under Mike's supervision, but not his last. Within the year, he was the co-supervisor with David Suzuki of a young masters student named Caroline Astell.[35] Astell, a Vancouver local and UBC graduate, was excited about the new advances in molecular biology and later joined Mike for her doctoral studies. Both Wilson and Astell eventually had academic careers thanks to their graduate work with Mike, who gained a reputation for being willing to work with female students.

Mike's enthusiasm for basic biochemical research and his willingness to work with graduate students caught the attention of Marvin Darrach, Head of UBC's Department of Biochemistry. In 1964, Darrach suggested appointing Mike as an Associate Professor in his department. Mike expected to become an "honorary" professor following standard practice but Darrach wanted more. To Darrach, this outstanding young scientist already had an international reputation in nucleic acid chemistry and had already been a great help to both students and faculty in his department. Because he was the kind of scientist who would attract high quality students, Mike was offered the responsibilities of a regular Associate Professor, although part-time and without salary. Mike soon joined the department with the consent of his boss, Hugh Tarr, who wrote to the appropriate Dean to approve the appointment.[36]

Another UBC department Head, Bill Hoar of the Department of Zoology, wanted to appoint Mike as an honorary professor and asked his Dean for permission. Hoar explained that Mike was already active in the department advising students, participating in seminars, providing hormones for analysis, and cooperating with the department's own fisheries research.[37] Mike happily accepted the honorary position and took a great interest in the Zoology Department, where, the following year, he attended a series of ad hoc lectures provided by David Suzuki. He was impressed with Suzuki who eloquently described the key papers appearing in such leading journals as *Genetics, Science, Nature,* and *Journal of Biological Chemistry*. Mike and his students participated freely and enthusiastically in the discussions.

By mid-1964 Mike effectively had three jobs: his research according to the mandate of the Fisheries Research Board at the Vancouver Technological Station, his independent, grant-supported research, and his unpaid work as a UBC "professor" that included department meetings, seminars, and graduate student advising. To Mike, there seemed to be nothing exceptional about having UBC students work with FRB scientists. Hugh Tarr was an Honorary Lecturer in the Department of Zoology, and Dave Idler had earlier provided graduate supervision for a masters thesis on fish steroids. Besides, the new Board Chairman in 1964, F. Ronald Hayes, was a strong supporter and advocate of FRB/university liaison more generally.[38] Mike worked long hours into the evening and on weekends to keep up with the demands, finding the work enjoyable and satisfying. His work was well regarded by colleagues, published in the Board's own *Journal of the Fisheries Research Board of Canada* and in the *Canadian Journal of Zoology*, the *Journal of the American Chemical Society*, the *Journal of Molecular Biology*, the *Canadian Journal of Biochemistry*, and a short note in the prestigious journal *Nature*. In addition, he was on the executive of the Vancouver Section of the Chemical Institute of Canada from 1963 to 1965. Colleagues at the Technological Station saw him as a dedicated researcher who seemed to live and breathe science but little else. To them, he was a very pleasant person but seemed a little single-minded in his interests; they wondered if he had a life outside of the lab.

Of course, he did, although he now had much less time for his other interests than he had when he was a post-doctoral fellow. Upon their return to Vancouver, Mike and Helen had set up home in a rented house not far from the university. Their first child, Tom, was born in 1961 followed by a second son, Ian, in 1963 and a daughter, Wendy, in 1964. Like his own father, however, Mike left child raising and household management to his wife, who was a devoted mother. In 1964, Mike and Helen bought a house with a view of the ocean and mountains in the neighbourhood of West Point Grey, a short bicycle ride from the FRB laboratories and the university. Mike was not much of a handyman at home but he eventually built a sandbox for the children and enjoyed mowing the lawn into the striped pattern popular in England. The Smiths were far from wealthy so they made do with spartan decor, although Mike would have bought art (especially west coast Native designs) if he had had the money. On Sundays or in the evenings, Mike found a little time for hiking, skiing, classical music concerts, and hosting the occasional home party. His parents visited every couple of years and sometimes joined the family for a vacation, travelling with Mike, Helen, and their children in the Volkswagen station wagon that had replaced the sports car.[39]

Quite often, Mike's social activities and his work were combined. One year, for example, he built a small sailboat at home with the help of Gordon Tener who, as a faculty member in the UBC Department of Biochemistry, had a growing reputation of his own in nucleic acid research. As they worked in the basement during the evenings they discussed the chemistry of nucleotides. The boat was completed in the summer of 1963, christened with champagne, and sailed in Burrard Inlet on a number of occasions. Mike proudly boasted that the boat did not sink, but eventually it became a planter in the backyard of Mike and Helen's home. The important part of the project, however, was meeting with a colleague to discuss science, and although Mike would build no more sailboats, he would frequently meet with colleagues over a beer and talk about science or politics.[40]

Life in Vancouver was going well for Mike in many respects but he was unaware of a growing problem at the Technological Station. Mike's enthusiasm for his independent research and UBC-related work was not as well regarded as he thought. Superiors at the Fisheries Research Board in Ottawa were beginning to reconsider whether he was a suitable employee, and other scientists at the Technological Station in Vancouver felt upstaged. The first hint of a problem came in 1964 when Mike noted that his salary was considerably lower than a UBC biochemist's, and even lower than a Fisheries Research Board scientist's. This was due mainly to earlier readjustments of the FRB pay scales, a situation that bothered Mike but which he did not take personally. He wrote a memo to his supervisor, Hugh Tarr, stating "my salary is about $1,000 out of line." He was therefore very interested in the prospect of an honorarium from the Department of Biochemistry but FRB Chairman Hayes in Ottawa enforced policy and forbade him from accepting. Mike thought his productivity was higher than a typical Board scientist and argued that his UBC involvement in no way compromised his fisheries research. Two days before Christmas in 1964 he asked Tarr for a raise, threatening to accept a UBC honorarium if the increase was not forthcoming. Tarr did not respond, leaving Mike with neither a raise nor an honorarium from the university.[41]

There were other clues that working relations between Mike and his boss were strained. A rumour circulated that Mike had violated protocol by borrowing an FRB truck for a field trip to an Oregon fish hatchery without completing the appropriate paperwork, and another claimed that he lent an FRB microscope to David Suzuki without proper approval. Mike sometimes complained about the "government mentality" at his workplace and often showed little concern for paperwork or protocol. Not everyone believed the stories about the truck and the microscope, but something was causing tensions. Mike's students thought that the environment at the Technological Station was rather unfriendly.

Suspicions of growing antagonism were confirmed in July 1965 when Mike realized that he had not received his annual salary increase. FRB staff expected the laboratory Director to recommend their raise as part of the regular routine. Mike was so incensed that he wrote to the Fisheries Research Board Assistant Chairman in Ottawa to protest, noting that his salary had fallen behind for bureaucratic reasons and asking whether the quality of his work or his university involvement was at issue. As far as Mike was concerned, both were beyond reproach. He asserted that his research in nucleic acids was relevant to the metabolism and even flavour of commercial species while his salmon studies were as important to the health of British Columbia stocks as medical research in the last century had been for human health. Mike insisted that his UBC collaboration was consistent with Board policy even if others at the Vancouver station were apathetic or hostile toward the university. He enjoyed supervising students and did high-quality research for the Board. Satisfied that he had a compelling case, Mike concluded that a salary increase was highly appropriate. He admitted frustration with FRB bureaucracy and the frequent need for Ottawa's approval, but he was optimistic about an exciting career at the Vancouver Technological Station.[42]

While waiting for a reply, Mike asked Tarr to confirm Board approval of his plans to supervise students in nucleic acid research. Mike had a new post-doctoral fellow, Robert Klett, and told his boss that fellows were necessary for a good laboratory and to provide the scientific opportunity currently lacking in Canada. Tarr did not respond to the memo and Mike soon knew why. Board Chairman Hayes telephoned Mike directly from Ottawa to issue an ultimatum: "settle differences with Dr. Tarr or lose job." Mike was incredulous and continued to defend his position, despite Hayes' warning. Once more Mike described to Tarr his work with Gordon Dixon in the Department of Biochemistry on salmon pituitaries and his work with David Suzuki in the Department of Zoology on crab DNA, asking for approval to continue. This time there was a response. Tarr, who claimed to be relaying the instructions of Hayes, refused to allow the projects and ordered Mike's technician out of the FRB laboratory.[43]

Mike was shocked. Either he would have to curtail his activities or he would have to work elsewhere, and neither option was attractive. Some people blamed Tarr for the hostility, believing that he regarded Mike's research as too esoteric. Tarr was accustomed to applying his research to commercial ends, having earlier developed a highly successful method of storing fish using refrigerated sea water and a method of preserving fish using tetracycline. His promotion to management of the Vancouver Station was prompted in part by industry complaints about his predecessor. While Tarr maintained significant research activity, especially in the fields of microbiology and food technology, FRB administrators expected he would be a little more responsive to the wishes of industry. Although Tarr had a personal interest in basic research and gave his scientists considerable freedom, he enforced the FRB mandate to keep research relevant to fisheries and oceans.

Tarr's disapproval of Mike's activities puzzled friends who regarded the Director, a Fellow of the Royal Society of Canada, as a strong research scientist who was very interested in such fundamental questions as the properties of DNA. Tarr had hired Mike in full knowledge of the young scientist's interests and background, provided him with a short leave to an American laboratory in 1964, and cooperated at least once with him to sponsor a visiting scientist. As for university involvement, Tarr had an honorary appointment in the Department of Zoology and had been consulted before Mike joined the Biochemistry Department. Tarr provided "full support of these proposals."[44] Tarr had encouraged Mike's diverse interests, and it seemed unlikely that the veteran FRB scientist would disapprove of Mike's work simply because of its more fundamental nature.

Some of the antagonism to Mike may have come from administrators higher in the bureaucracy who disapproved of overly independent FRB scientists. The Fisheries Research Board's central administration since 1953 had become more centralized and authoritarian, intent on controlling policy for all the formerly independent laboratories across the country. Ronald Hayes, the Board's Chairman after 1964, reinforced the tendency toward autocratic leadership. The Board Chairman could make life unpleasant for laboratory Directors who failed to follow central policy. The Director of the Halifax Laboratory had resigned a few years earlier over the issue of local authority, and Peter Larkin, the independent-minded zoologist and Director of the Nanaimo Research Station, resigned in 1966 after confronting Hayes. Larkin returned to UBC and left Board colleagues to "the battle against the elephants."[45] Hugh Tarr was five years away from his retirement and not likely to challenge the central authority; he generally followed protocol and sought approval from Ottawa for his decisions. If Mike was out of line with Board policy it was Tarr's job to enforce compliance or find a replacement.

In fact, Mike *was* challenging Board policy. Not only had he demanded permission to receive a UBC honorarium but in the late spring or early summer of 1965, he had sent a memo to Ottawa suggesting conditions for university/Board cooperation. Hayes, responding to Hugh Tarr, agreed that research problems should be designed for the benefit of the student rather than for economic utility, but only for problems within the scope of Board interest. He added that post-doctoral fellows should be chosen by the National Research Council and graduate students should be chosen by university administrators or laboratory Directors; FRB staff members were not to select their own students. Hayes further declared that graduate and post-doctoral students probably should not work in the same laboratory. Mike did not agree. He wanted the freedom to choose his projects on academic grounds, to participate in the selection of his assistants, and even to hand-pick his students, which he had already done. His post-doctoral fellow had a United States National Institutes of Health fellowship, not one from the National Research Council. Mike also wanted to publish where he saw fit and was no doubt annoyed by the need for permission to publish outside the FRB's own journal.[46] Unfortunately for Mike, he was no longer in Khorana's lab at the B.C. Research Council supported by Gordon Shrum and NRC President Edgar Steacie.

Other staff at the Vancouver Station may also have influenced Tarr. Mike was feeling quite confident of his abilities as a scientist but he was not always impressed with the work of his immediate and well-established colleagues. (He may have had reason to feel superior. When challenged by the editor of the *Canadian Journal of Biochemistry* about a submission, Mike insisted he was right about the adequacy of his data and the proof lay in his eight years of experience studying nucleic acids, five of which were with Khorana. Anyone experienced in the field, he stated bluntly, would understand his procedures.) In return, his FRB colleagues seemed to resent this young upstart who worked evenings and weekends with graduate students (his own and those of others), undergraduate summer students, a post-doctoral fellow, and university professors, including David Suzuki who was using Mike's laboratory to learn biochemistry. Mike was playing an active role in the life of the Departments of Biochemistry and Zoology and using laboratory space and supplies (including young salmonids) at the Technological Station for the UBC projects. Mike saw himself as a leader in establishing highly productive scientific exchanges between the university and the Fisheries Research Board, but others saw him outperforming Technological Station scientists who were increasingly annoyed. Some FRB administrators even dismissed Mike's skill as a scientist and thought him presumptuous. Internal rivalries with FRB colleagues may also have pressured Tarr to maintain the peace at Mike's expense.[47]

Mike was not fired outright but the message to conform to Board expectations or leave was certainly explicit. In his good-natured determination to pursue what he thought were interesting and important research questions, Mike had not noticed that he was aggravating others. Always shy of direct confrontation, he was terribly upset upon receiving the final ultimatum that struck at some of his deep-seated insecurities, surprising many of his friends and colleagues in Vancouver who had never seen him so distressed. He was not ready to abandon his research ambitions, but he could not continue where he was. Fortunately, Mike had influential friends who were dismayed at the prospect of losing a brilliant researcher and launched a plan to get him a new job. Several members of UBC's Department of Biochemistry, including Gordon Tener, were Research Associates of the Medical Research Council of Canada. If the university did not have the funds to hire Mike, perhaps the MRC did.

The Medical Research Council was a relatively new national organization created from the National Research Council in 1960 to fund health research when government support for health services increased following the Second World War. In 1957, a federal/provincial hospital insurance plan had been introduced as Canada moved toward its comprehensive, universal healthcare program supported by federal and provincial funds. A new body separate from the NRC was needed to administer the increasing amount of money set aside for health research. When the federal government announced its commitment to universal medical insurance in 1965, medical researchers across the country, including members of the UBC Department of Biochemistry, lobbied successfully for additional research funding. The MRC budget increased fifteen-fold during the 1960s.[48]

The MRC Research Associate program was intended to increase the number of medical researchers in the country by providing competitive salaries to promising young scientists of demonstrated ability who were employed in basic or clinical science departments of university medical schools. Research Associates as defined by the MRC were independent scientists who received an initial grant to begin their research programs, but were subsequently required to write funding proposals for peer adjudication. UBC's fledgling medical school (first admitting students in 1950) was beginning to grow, but the basic science departments like biochemistry were small and looking for opportunities to hire more faculty members.[49] Mike seemed like an ideal candidate for an MRC Research Associate position.

The Head of the Department of Biochemistry, Marvin Darrach, was familiar with procedures at the MRC. Having been on the committee to appoint Research Associates, he knew the right people and proper protocol. Because recommendations for MRC Research Associates required approval by the sponsoring university's president, Darrach and John McCreary, Dean of Medicine, sent a letter in September 1965 to UBC's new President, John Macdonald.[50] Macdonald had initiated sweeping reforms of higher education in the province by recommending several new post-secondary institutions and reorienting UBC as the province's principal research university. Among the visions he had for UBC was an increase in science and science-based professional education, claiming that "one Enrico Fermi is more valuable than a thousand ordinary PhDs." Macdonald, a microbiologist and dental scientist who wanted dedicated research scientists at his institution, willingly signed his name to Mike's application and sent the request to Ottawa. Khorana, well established as an intellectual powerhouse, also wrote a letter to support Mike's application.[51]

Although a decision by the MRC took several months, Mike was confident of success. He wrote to the Council a month after his application had been submitted requesting $43,450 for moving expenses and new laboratory equipment. This was a large sum of money, he admitted, but it would be expensive to move all his work in progress and begin new studies. In January of 1966 it was still undecided if he had the position; the MRC requested a transcript from the University of Manchester which Mike immediately arranged. Finally in March, the news was out that he would be appointed a Research Associate of the Medical Research Council of Canada at a salary of $13,600 per year, if he chose to accept.[52]

Mike was very happy to accept. His FRB salary for 1965 had been only $11,200, so he would receive an increase in pay. More importantly, he could now dedicate himself to his basic research. Because he still needed a full-time university position, Darrach recommended an appointment to Dean McCreary, reminding him that Smith was an outstanding scholar and an enthusiastic and dedicated teacher. Two professors in the department, Gordon Dixon and Gordon Tener, added letters of support. So too did Hugh Tarr, who may not have felt any personal animosity.[53] Perhaps Tarr realized that the FRB was simply the wrong institution for such a determined and meticulous researcher intent on exploring fundamental questions of biochemistry and molecular biology. McCreary approved Mike's appointment as an Associate Professor and congratulated him on the MRC award, noting that the position provided the maximum degree of security possible for the best researchers in Canada.[54] With a UBC commencement date set at July 1, 1966, Mike resigned from the Fisheries Research Board with a pleasant note to Tarr, copied to Chairman Hayes and others. Not one to bear a grudge, Mike tried to assuage any lingering ill-will and avoid future conflict by recalling his exciting research from the previous few years and promising to continue with some of his earlier studies, perhaps in cooperation with the FRB. He felt secure enough about his future at UBC to decline an invitation from York University to apply for the position of Head of the new Department of Biology, and spent May and June moving his laboratory equipment to UBC.[55]

Given the lacklustre institutional support he had received while with the FRB, Mike may not have been aware how important he was to the Department of Biochemistry. Darrach thanked the MRC for providing the funds to keep this talented scientist not only in Canada but in his department. Universities in the United States had considerably more money to attract scientists like Mike, and other departments at UBC were already interested in him. Darrach requested that Mike's research grant be made available immediately so as not to interrupt his work, explaining to the MRC how modest Smith's financial request was given the size of his research staff — Mike had already accepted another graduate student — and the vigorous and expensive nature of his work. After searching for departmental funding to cover some initial costs, Darrach petitioned his Dean for additional money to cover moving expenses and alterations to the department's facilities. Particularly urgent was plumbing for fish tanks.[56]

To the extent that it was possible at UBC in the 1960s, Darrach rolled out the red carpet for his new faculty member. Although the MRC stipulated that their Research Associates should spend seventy-five percent of their work on research, Mike agreed with the other Associates in the department that they should be full participants in the life of the university and accept teaching and service duties. The terms of the Research Associateship forbade him from holding administrative office or receiving remuneration for outside work, but these restrictions were perfectly acceptable to Mike. He now had exactly what he wanted: freedom to pursue his research interests, a home in a magnificent setting, and good friends and colleagues. For Mike, it was now time to have some "real fun."

3

HAVING FUN

When Mike became a Research Associate with the Medical Research Council, he was already working long hours to pursue unique lines of inquiry and publish new insights or discoveries, but now he could devote himself to basic research that was guided by his own curiosity and the intellectual development of his academic field. The MRC would review his progress every five years, beginning in 1968, but he was free to conduct his laboratory in whatever way he saw fit. For his laboratory to be productive, Mike would need certain technical, social, and political resources. He had to keep current with recent research in biochemistry and molecular biology through considerable reading, numerous out-of-town conferences, and the occasional leave to study in other laboratories. He needed grants from the MRC or other funding agencies for expensive supplies and to provide salaries for professional assistants and technicians. He had to recruit competent students, post-doctoral fellows, research associates, and technicians. Some of them might need a small stipend although post-doctoral fellows usually came with their own scholarships. Mike also worked best when relations with colleagues and associates were congenial and in an environment unencumbered by bureaucratic restraints. To many people, establishing a research laboratory would be daunting, but to Mike, with his immense curiosity and tenacious drive to answer fundamental scientific questions, running his own laboratory allowed him to work at what he loved.

Mike's first job in the summer of 1966 was to establish his laboratory space, a task made easier by the generous start-up fund of $100,000 made available by the MRC.[1] His graduate students, a couple of undergraduate summer students, a post-doctoral fellow, and a technician were already part of his team. He was given a basement room in the Medical Sciences Block "A" Building (later renamed the Copp Building), home of the Department of Biochemistry at UBC in the new medical sciences complex across from the War Memorial Gymnasium and the Empire Swimming Pool. His room was small (about 450 square feet), mostly below ground level, and had only one row of small windows near the ceiling. It was a little dingy but served its purpose for many years by providing such basic features as work benches, racks to store chemicals, sinks, gas nozzles for bunsen burners, and a fume hood for the more noxious chemicals. Mike put his fish tanks — he planned to continue his salmon studies — in a windowless room across the hall. His office was located several floors above and because it shared a telephone line with the laboratory, a buzzer was installed for Mike to alert lab members of an incoming call for them and vice versa. It was not an ideal location, but great things could be done under poor conditions.

Mike was fortunate that the university was beginning to grow and could provide him with any laboratory space at all. A few years earlier, before construction of the new Medical Sciences Buildings, the Department of Biochemistry had been housed in old army huts that had no room for another laboratory. UBC's Faculty of Medicine, the administrative home of the department, was entering a period of growth that paralleled developments in medical research and education across the country. UBC had provided degree programs in nursing and public health in the Faculty of Applied Science since the 1920s, but it had taken university administrators nearly thirty more years to convince the provincial government, local physicians, and philanthropists of the importance of a medical school in British Columbia. As government funding for the university increased during the 1960s, local philanthropists also contributed large sums. In particular, the Koerners, Czech immigrants who had profited handsomely in the lumber industry, and the Woodwards, pioneer Vancouver retailers, helped build UBC's new Medical Sciences Centre that included research laboratories, lecture halls, a biomedical library, and, eventually, a teaching hospital.[2]

As a Research Associate of the Medical Research Council of Canada in the Faculty of Medicine, Mike argued that his marine research explored basic biochemical problems that had wide-ranging implications for medicine, and he criticized scientists who were "seeking answers to questions of only limited significance." The MRC agreed and Mike's grant proposals rarely failed. New university professors often continue working on projects that have already been successful and Mike continued with his marine interests for another ten years, isolating hormones and other natural products from marine organisms and exploring the physiology and biochemistry of the sexual maturation of salmon and trout. These research areas fit with the classical biochemical concerns that accompanied the rise of scientific medicine, including metabolism, nutrition, and the physiological role of proteins, but also extended to understanding macromolecules such as DNA. In keeping with his training as a chemist, Mike championed a view of biology that explained life processes in terms of chemical processes, a view that was gaining preference in scientific circles worldwide. His other line of inquiry, the chemical synthesis of oligonucleotides, explored various properties of the synthesized molecules. Because the basic principles of molecular biology (what James Watson called the "central dogma") had recently been established, Mike's work explored important problems in nucleic acid chemistry within these new theoretical boundaries.[3]

The "central dogma" of molecular biology
is summarized in the diagram

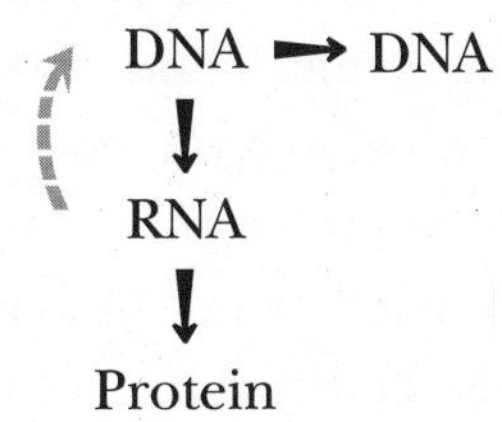

The diagram shows DNA doing two things: replicating to make more DNA and being copied into RNA, thus transferring the genetic information into a messenger molecule. This messenger RNA is decoded by the process of protein synthesis, making a protein whose amino acid sequence is specified by the precise order of the nucleotides in the RNA. In addition, some viruses copy the RNA into DNA (e.g. retroviruses, including the virus responsible for AIDS). In yet other RNA viruses, a special RNA polymerase copies RNA into more RNA.

Almost immediately after his appointment to UBC, Mike was enjoying his new freedom to work with the students and fellows of his choosing. One of his first projects was to characterize unusual nucleotides in the DNA of small marine crustaceans called copepods, which were collected from the ocean. To this end, he arranged one of his periodic trips on a research ship, the *Ekholi,* on a sunny, warm day in August 1966 and brought along several members of his laboratory. The ship travelled past the Port of Vancouver and up Burrard Inlet, sweeping the sea several hundred feet below the surface with a large canvas sack that funnelled the copepods and other small marine organisms into an attached metal canister. Mike enjoyed the sunshine and the good company until suddenly, tragedy struck. Sherri Hepner, who had just joined Mike's lab as a graduate student, had a known cardiac defect that unexpectedly caused her heart to fail. Frantic attempts were made to revive her as the ship rushed to medical personnel waiting at Deep Cove, but despite their efforts she died before reaching shore. Mike was devastated and for some time a gloomy atmosphere pervaded his laboratory.

Work slowly returned to normal as Mike and his students overcame their grief and settled into regular routines. University research laboratories are typically organized under the supervision of a single professor who guides the investigations and controls budgets, personnel, and the arrangement of equipment. Laboratories "belong" to the supervising professor, and those who work in them are members of a research team. As a university professor, Mike was no different. He chose his researchers carefully, contacting friends in other institutions to screen for bright and congenial associates, especially the post-doctoral fellows whom he preferred over doctoral students.[4] He had an eye for spotting talent. His lab soon acquired a distinctive characteristic: more women worked in his lab than was typical, perhaps because Mike was approachable and more willing to work with females than were other scientists in his field. In many ways he was a casual supervisor who never worried about dress codes, regular progress reports, or other formalities, and the lab was a pleasant and friendly work site. He procrastinated with administrative chores, putting off paperwork to the last minute, but he was certainly not casual when it came to his research. He was a persistent, dedicated, exacting, and sometimes impatient scientist who worked long hours and expected the same from his laboratory colleagues. Mike impressed upon his students the importance of carefully writing up all experiments, successful or not, and always maintained that you could learn something useful from a failed experiment. (He did not always heed his own advice — his notebooks were often full of casual and cryptic descriptions of experimental details.)

Mike had to drive himself. The field was competitive and the laboratory had to publish "firsts" to achieve peer recognition, funding, and the excitement of a new discovery. He usually started by eight o'clock in the morning and frequently worked into the evening, circulating regularly through the laboratory, chatting with students or fellows, and holding meetings from time to time when it seemed necessary. He was not as "hands-on" as he would have liked, but he did have the broader view of what needed to be done. Doctoral students were often treated as post-doctoral fellows, which meant they had to learn quickly the practices and procedures that would ensure success as Mike rarely designed experiments or provided step-by-step guidance. What he did provide were problems to be solved or questions to be answered in the areas of nucleotide synthesis, hormone isolation in marine animals, or the sexual maturation of salmon. Not all graduate students were ready for such independence and some withdrew from the program or found other advisors. Some post-doctoral fellows also shortened their stay because they did not have the guidance they expected. Yet Mike was always willing to acknowledge their contributions in the laboratory. Everyone who helped with an experiment was entitled to recognition in the ensuing paper; even Sherri Hepner received posthumous credit.

Mike also worked Saturdays, but not as intensively. Helen generally took care of household duties (that was Helen's job, Mike once told a friend) but on the weekend the young professor and father of three sometimes brought his children, Tom, Ian, and Wendy, to the lab. They often amused themselves building complicated structures from green rubber stoppers or blowing bubbles with the pipette washer. The vortex, a small vibrating device for mixing liquids in test tubes that was activated by touch, proved to be the most delightful toy. Even Fisher Price could not compete! Mike usually took his children home around noon and returned to play field hockey at the south campus in the afternoon. Sundays were often free of laboratory work, and for many years Mike found a little time for skiing, camping, and fishing trips with his family and as his children grew he occasionally drove them to soccer, baseball, and swimming practices, attended public school meetings and birthday parties, and visited out-of-town relatives. But gradually he found himself spending less time with his family and more time at the lab.

Although Mike wanted results from his laboratory, he knew from personal experience that scientific experiments did not always work as anticipated. He also knew from his own graduate work that students, like others, sometimes encountered unexpected barriers. It was important to learn from mistakes or dead-end lines of investigation and start anew, particularly since some of Mike's vast supply of intriguing ideas were unsuccessful. Caroline Astell, who transferred to study under Mike after completing her masters degree in David Suzuki's laboratory, learned this the hard way. Her ambitious project grew out of her earlier work with Mike while he was with the Fisheries Research Board. She was to identify the distribution of guanine (G) and cytosine (C), two of the four bases in DNA, in a special form of DNA found in crabs. Mike suggested a recently published method to establish whether these bases were clustered in discrete regions, interspersed at regular intervals, or interspersed randomly. The first step was to attach methyl groups to the G, which would render the sugar/phosphate backbone of the DNA susceptible to cleavage with alkaline solutions. Cutting the DNA at G residues would, Mike reasoned, leave chains of alternating A/T nucleotides the length of which would reveal the dispersion pattern of GC base pairs. After two and a half years of trying unsuccessfully to methylate the G specifically, Astell realized that the method would not work.[5]

Astell told her supervisor, who agreed with the verdict and set her on another project to synthesize oligonucleotides and test how well they bonded to others under different temperatures. Mike was curious about how well the two complementary strands of DNA stuck to each other. (He then called it quits on crab DNA studies, admitting that "marine crabs are not the most convenient laboratory animal.")[6] The task he set before Astell was very time consuming. Beginning with large supplies of purified nucleotides, she first synthesized the oligonucleotides using the phosphodiester method developed in Khorana's lab ten years earlier. Astell had to do the steps manually as Mike's lab could not afford the proper equipment for gathering or analyzing the fractions automatically. (Because she had to use large, plastic buckets to hold the salt solutions used in washing away impurities, the lab joked that she was getting her PhD in bucket biochemistry. She even developed a sensitivity to some of the chemicals, as did an unwitting summer student.)

The phosphodiester method was still inefficient — at best a twenty-five percent yield of the starting material — which meant that it would easily take two or three months to make an oligomer of nine units in length.

The second step of Astell's project was to test the thermal stability of bound oligonucleotides, a relatively rapid and straightforward procedure. She attached the synthesized oligos firmly to cellulose paper using a relatively benign reagent, the efficiency of which could be doubled with a second, overnight application. The oligo/cellulose mixture was stirred vigorously and packed into a small, glass column surrounded with a water jacket. Complementary oligomers were allowed to bind to those in the cellulose slurry and the column washed with salt solution to remove the unbound oligomers. Astell measured the strength of the binding between slurry-oligomers and their complements by gradually increasing the temperature of the column while pouring salt solution through the column. A series of experiments showed that the temperature at which an oligomer bound to another depended on its length, the number of GC base pairs, and whether one of the oligomers was a ribo-oligomer (RNA) rather than a deoxyribo-oligomer (DNA). (RNA/DNA interactions were less stable than DNA/DNA interactions.) Other studies showed that the oligomers could also bind to each other if there was a mismatched base pair near the centre of the synthetic piece, although such interactions were less stable.

These studies were highly successful and Astell finished her PhD in 1970, subsequently co-authoring four published papers with her supervisor. This was basic, curiosity-driven work that Mike thought could be useful in identifying naturally occurring genes. Genes were now known to be the segments of DNA that encoded particular proteins, and Mike thought that a synthetic oligomer might be a useful probe to find genes in living organisms. The study also suggested a possible method of purifying certain nucleic acids (messenger RNA) but otherwise it had no immediate, practical application. It would later provide a basis for a new scientific breakthrough, but this was not foreseen at the time.

Although Mike expected colleagues to work hard, he was fun

to be with — very enthusiastic, lively, and interested in everyone's research, especially when the projects were going well. He was available to members of his laboratory, providing general advice or suggesting resources even if he did not plan each individual experiment. He loved to talk science and politics and his gregarious nature ensured that there was a steady flow of visitors through the laboratory. He was typically blunt and outspoken with his opinions, but friends interpreted his candour as brusque honesty rather than rudeness — typical Lancashire frankness. Mike's earthy sensibilities and enthusiasm for science and new discoveries led to his reputation as a "nature reader of *Nature*," meaning that he usually headed to the men's washroom for some peace and quiet to read new issues of the journal *Nature*, a practice that was repeated at home with newspapers. Mike later joked that the men's room should be recognized as the place where he had his best ideas.[7]

In addition to his role as scientific leader, Mike also assumed a role as social leader in the tightly knit scientific community. Congenial company and social approval supported his academic work. Field trips to collect marine specimens provided one sort of social activity, sending Mike and his colleagues to the shores of Stanley Park (sometimes late at night), Washington State, or other sites along the west coast. Some trips took several days and lab members even went on their own camping trips without Mike just for fun. Other trips were to attend seminars or conferences. One took several carloads of UBC professors, fellows, and students to Seattle, Washington, to hear a leading figure of molecular biology, Jacob Monod, speak on allosteric enzymes. Passengers in the convoy took great delight in confounding the American border officer with the diverse collection of nationalities and the esoteric purpose of their trip. When the third car arrived at the border, the officer asked them, "Just what in the hell is an allosteric enzyme?" Whether he was enlightened or not, all of the UBC scientists had a good laugh.[8]

Parties in the laboratory were also common — just about anything could be celebrated with a lab party, especially a successful experiment. Mike liked popping the cork from a bottle of inexpensive champagne, acting like a little boy opening a birthday present. At one of the Christmas parties Nadine Wilson produced an incredible recipe for eggnog which was prepared in the cold-room facility in large laboratory beakers. The eggnog passed from the beakers into a glass cylinder (a 5cm Kontes column, 100 cm in length) via rubber tubing and then into individual glasses held below. The set-up looked like just another piece of laboratory equipment, a Sephadex column. Marvin Darrach, the Department Head, interrupted the fun when he dropped by and discovered the eggnog, but seemed impressed with the young scientist's ingenuity and the festivities continued.

Mike's role as social leader went beyond field trips and laboratory parties when he invited friends, colleagues, students, and neighbours home for fun and games in the evening. These soirées were usually announced to members of his lab at the last minute and although Mike's wife, Helen, sometimes seemed a little overwhelmed by the mini-invasion of twenty to thirty people, she worked hard to plan and organize an evening of lively parlour games. Teams of two competed to identify magazine clippings, test their memory with "Kim's game," and jump on a Pogo stick in the basement while larger teams later competed to pass an orange from one person to the next, holding the orange under their chins. The beer bottle extension challenge in the living room was particularly difficult. Contestants held an empty beer bottle in each hand and leaned onto the floor, reaching forward as far as possible so that much of their body weight rested on the bottles. The contestant then transferred both hands to one bottle and reversed the extension until he or she could stand up without touching the floor. Some were so hopeless at this game they did not try, but both Mike and Helen were very adept. Helen was particularly skilful thanks to years of yoga training that endowed her with remarkable athletic ability and flexibility. By the time the games were over, everyone was close to exhaustion and turned to eating and drinking. The noise of the party often woke Mike's and Helen's children who would appear in pyjamas and join the adults.

In the winter, Mike's parties included tobogganing on the steep roadway next to his house. Snow was not common in Vancouver, but when it did fall the road was closed and locals spent hours sliding down the thick, wet slush. The Smith household became known as a place of hospitality where a mug of mulled wine or a bowl of chili could be had on a winter night.

Good rapport with faculty colleagues continued to be important to Mike. His enthusiasm for his work seemed almost to depend on cordial relations, and many of the professors in his department came to his house parties. He spent Friday afternoons with a small group of close friends drinking beer under the stairs in the Faculty Club and chatting, a practice that became a lifelong ritual. They usually observed a one-beer limit while they shared information about science and institutional politics. Many of these friends were influential administrators, senior scientists, and rising young academics who formed a powerful if informal social network. At first Mike could not enter the Faculty Club because he refused to wear a tie, but capitulated when Helen bought him one that bore the word "bullshit." He was still a rough diamond who delighted in barroom ballads, the English television comedy *Monty Python's Flying Circus,* and spelling rude words in Scrabble, but to many colleagues in the biochemistry community he had "the enthusiasm, energy, and goodwill which make collaboration a pleasure." A colleague at another university claimed, "I cannot think of a person that I would prefer to have as a scientific companion or personal friend." Mike came to be well-known and generally well-liked in North American scientific circles, including the Canadian Biochemical Society, the American Society of Biological Chemists, and as Chairman of the Molecular Biology Program Subcommittee of the International Congress of Biochemistry from 1967 to 1979.[9]

Mike suffered when collegial relationships faltered, as happened with Gordon Dixon. A couple of years earlier, while still with the Fisheries Research Board, Mike suggested to Dixon that they collaborate on studies of the natural synthesis of basic nuclear proteins (specifically protamines) in salmon spermatozoa. Although the collaboration was scientifically productive, they had very different backgrounds and "Old Country" sensibilities. Mike was the gregarious and earthy market gardener's son with a degree from Manchester University while Dixon was the serious and more independent upper middle-class professional's son with a Cambridge degree. Dixon was well established as a scientist, having achieved considerable acclaim with his work on proteins since arriving at UBC in 1964. Among his accomplishments, he separated the two protein chains of bovine insulin, inactivated the protein, and then developed a procedure to recombine the chains efficiently to produce a biologically active hormone. For reasons Mike could not explain, their partnership came abruptly to an end after the publication of a note and a review in 1966 and 1968. Mike complained to the MRC that Dixon had taken charge of the project and severed the partnership suddenly and unexpectedly. Even one of Dixon's PhD students was annoyed with his supervisor's behaviour. Feeling discouraged and a little betrayed, Mike avoided new research on salmon spermatozoa and shifted his attention to nucleotide synthesis.[10]

Except for minor setbacks, Mike's lab was generating important results. He was willing to share information, but not at the expense of losing a publication in the "publish or perish" environment of the university. From 1966 to 1969 Mike and his laboratory published some fourteen papers; from 1970 to 1974, they published another fourteen. This was not a particularly high number but they were generally well regarded and the lab was still small.[11] His marine work — identifying hormones in fish pituitaries or unusual DNA in crabs, separating cell types in salmon testes, and characterizing unusual nucleotides in a marine copepod — provided findings of a broad and general interest to biologists. His studies of the physiology and biochemistry of the sexual maturation of salmon and trout impressed colleagues by illuminating aspects of some key biochemical issues of the day, including DNA replication, protein biosynthesis, and the effects of hormones on the development of an organism. However, his greatest interest remained the synthesis of oligonucleotides and the characterization of their properties.

Although Mike was pleased with the studies in his lab, by the early 1970s he was rethinking his research priorities. To gain more prominence as a scientist, Mike had to establish independence from previous mentors by undertaking new, riskier projects. He wondered whether oligomers would truly be useful in isolating genes or messenger RNAs and, if they were, where these studies would lead. He considered changing his research focus from DNA to membrane proteins, a new and increasingly popular area of biochemistry. Membranes are the lipid (fatty) envelopes that surround cells and the many internal organelles within cells (such as the nucleus, the mitochondria, and endoplasmic reticulum).[12] Inserted into these membranes are specific proteins that facilitate the uptake or extrusion of molecules and transfer signals from the environment into cells. To introduce himself to some of the techniques unique to membrane biochemistry, Mike looked for an opportunity to study with a colleague already working in that area.

Such a colleague was not hard to find and in the spring of 1971, Mike had his first sabbatical leave to study at Rockefeller University. Sabbaticals are intended to allow one scientist to learn from another through a short apprenticeship and Mike welcomed the opportunity truly to get back "to the bench." The MRC granted him leave with full pay, an allowance for his research, a cost-of-living allowance, and a travel grant for his family. UBC also granted him sixty percent of his regular university stipend, but this was a very small amount since most of his salary came from the MRC. (Other academics, forced to rely only on sixty percent of their UBC salary, often could not afford a sabbatical.) Mike packed up his family and set off to New York to join the laboratory team of Edward Reich. For the next six months he helped purify and characterize the functions of novel membrane proteins from the electric organ of the electric eel. At the end of the six months, however, Mike returned to Vancouver and decided to continue his earlier studies of nucleic acid biochemistry. He resumed experiments on the interaction between complementary oligomers, with particular emphasis on mismatching oligomers, assisted by technician Patricia Jahnke and post-doctoral fellow Mick Doel.[13]

Colleagues were also visiting Mike for their sabbaticals, one in 1968 and two more in 1974. Mike even hosted a short visit from the renowned English molecular biologist Fred Sanger in 1973, who adjusted his tour of North America to include Vancouver. Sanger, who had won a Nobel Prize a decade earlier for the first sequence determination of a protein molecule (insulin), was interested in the new and more effective ways to synthesize oligonucleotides developed in Mike's lab by Shirley Gillam, a research associate who had recently arrived at UBC. At Mike's suggestion, she was exploring how the naturally occurring enzyme polynucleotide phosphorylase, which ordinarily degraded nucleic acids step-by-step, did the exact opposite in the presence of nucleoside diphosphates. Adding manganese ions allowed the enzyme to build DNA, step-by-step, much more quickly and cleanly than the chemical phosphodiester method. Sanger's group at the Medical Research Laboratories in Cambridge, England, needed a special oligonucleotide to help them develop a method of identifying the sequence of nucleotides in DNA, specifically the DNA of a small virus called phiX174. Mike willingly agreed to provide a few custom-made oligomers for his English colleague (and then arranged to take him salmon fishing aboard Gordon Tener's sailboat).[14]

All this laboratory activity was watched by the MRC with great approval. Mike wrote a report and received a site visit for his first preliminary review in 1968, learning a year later while at a conference in Atlantic City, New Jersey, of his five year reappointment at an annual salary of $16,000. The MRC was similarly impressed with the second review in 1973. UBC's Acting Head of Biochemistry at the time, Jim Polglase, provided strong support for Mike, describing him as a major asset to the department. More specifically, Polglase attested that "Dr. Smith has an exceptionally fertile imagination which generates unique and novel approaches to problems in biology." In addition to operating his own laboratory, Mike also organized local conferences, continued to donate time and money to the Biochemical Discussion Group, and generally contributed to the productivity and morale of the department.[15] Although Mike did not study problems of a distinctly medical nature, the MRC continued to accept his work as basic to the advancement of health.

Mike knew very well that all the productive fun he had in the laboratory depended on financial support from outside the university, especially from Canada's Medical Research Council. Although he was not an activist by temperament, he took very seriously his responsibility to ensure high standards and generous funding for the scientific community. In 1968 he joined a team of Canadian medical researchers who surveyed the field and proposed priorities and financial requirements for continued growth. Mike and the authors of the survey asserted that good Canadian medical research, funded by government via the MRC but conducted independently in universities, would enhance the quality of the nation's intellectual life. University research would also reverse the "brain drain" to the United States, provide a better society through better health, and ultimately help the economy. The Biochemistry Assessment Group of which Mike was a member criticized the current organization of Canadian biochemists (they argued that university departments should specialize — no department can excel in every area of biochemistry) and the quality of some of their studies. Above all, the Assessment Group insisted that Canadian biochemists must enter the international race for new, basic, scientific discoveries, supported financially by governments and universities and unencumbered by excessive teaching, administrative, or advising duties.

These efforts to secure funding paid off thanks to a favourable social, economic, and political climate. As Canada's publicly funded health systems grew, so too did government research funding, and in 1969 the MRC was legally reconstituted as an autonomous crown corporation. Mike then sat on the MRC grant committee from 1969 to 1973 to adjudicate funding requests. The MRC was proud of its peer-review process and Mike developed a reputation for shrewd judgment in identifying good proposals and rejecting poor ones.

Although Mike had been successful in establishing a productive laboratory and he enjoyed his research, there were aspects of his work at the university that he found discouraging. He had willingly accepted teaching responsibilities in several undergraduate courses offered by the Faculty of Medicine but was often frustrated by mixed reviews from undergraduates. The problem seemed to be that he taught biochemistry as a pure science and did not always illustrate the medical, agricultural, or biological application that interested students. Mike earned a reputation for showing too many complex illustrations too quickly, and his poor hearing, which he inherited from his mother, made it difficult for him to locate students when they asked questions in class. Students erroneously concluded that their instructor was uninterested in them. Other professors and graduate students thought that Mike was an organized, clear, and informative lecturer who provided the intellectual tools needed to solve biochemical problems and not just memorize and repeat information. Mike tried his best to please, but student criticism remained.[16]

Some of the complaints about Mike's teaching arose because of the diverse backgrounds of students in his course. Biochemistry courses at UBC were the responsibility of the Faculty of Medicine but the subject was first taught in the 1920s under the aegis of the Department of Poultry Science and then the Department of Dairying. (Agriculture professors still retained honorary appointments as lecturers in the UBC Department of Biochemistry in the 1960s.) In 1947, the Department of Chemistry tried to claim the field by offering to provide a single introductory course in biochemistry and by hiring Sid Zbarsky to teach it. When the newly established Faculty of Medicine began hiring staff in 1949, Zbarsky transferred to the new Department of Biochemistry.[17] Courses offered by the Department of Biochemistry thus served students in agriculture, chemistry, and medicine, as well as in zoology, microbiology, pharmacy, and nursing. Mike taught the course as "classical chemistry" and found it difficult to accommodate the students' varied academic backgrounds and educational objectives. By the late 1960s professors in the Faculty of Agricultural Sciences thought that the course was not appropriate for most of their students, and medical reviewers in the 1970s tended to give the department (and most of the Faculty of Medicine) mediocre teaching reviews.[18]

Mike was particularly annoyed with complaints from medical students who, in his view, sometimes asked spurious or irrelevant questions. He and his biochemistry colleagues felt that medical students were often too anxious to become physicians and did not fully appreciate the value of the basic sciences. Students once even successfully petitioned to have the biochemistry of nutrition removed from the curriculum. Some of the complaints, biochemistry professors insisted, could be attributed to worsening classroom conditions because faculty positions in the Department of Biochemistry were not increasing at the same pace as other medical departments at a time of escalating student enrolment. Class sizes were consequently high — medical students in 1974 were lost amongst the five hundred students in Biochemistry 410.[19]

Complaints from medical students hurt Mike's pride, but they also reinforced his suspicion that Faculty of Medicine administrators discriminated against his department in favour of the "clinical" departments. Not all physicians in British Columbia in the 1960s supported UBC's role in providing medical education, forcing the Dean of Medicine to curry favour with the profession lest universities outside the province or teaching hospitals assume the responsibility.[20] UBC's medical school also depended on government and public support for locally trained doctors and on favourable accreditation reviews by the American or Canadian Medical Associations. As a result, faculty administrators listened seriously to medical students.

Mike worried that the Department of Biochemistry, like the other basic science departments in the Faculty of Medicine (Physiology, Anatomy, Pharmacology) risked being overshadowed by the much larger clinical departments that were located closer to the Vancouver General Hospital some distance away. (Table 2 shows regular and clinical (on-site teaching) faculty appointments in two select departments.) Members of the Department of Biochemistry had long thought that their department was at a disadvantage in claiming resources from the faculty for space, new appointments, and staff support.[21] The comparatively low number of appointments in their department in the early 1970s suggested that this was still true, and the Department of Biochemistry could not obtain additional resources from the Faculties of Science or Agriculture.[22] Even Mike's research program was potentially at risk. He valued basic research, attracted science students working toward PhD degrees, and supported large research teams with post-doctoral fellows. Clinical departments, on the other hand, valued applied research, attracted medical students working toward MD degrees, and had independent investigators with graduate students.[23] The former emphasized laboratory experimentation, the latter application or professional practice. The rigid administrative hierarchy of the Faculty of Medicine provided little opportunity for ordinary professors like Mike to influence administrative decisions in their favour.

TABLE 2
Approx. Number of Faculty in Select Departments in the Faculty of Medicine

Department	1960	1965	1970	1975	1980
Dept. of Biochemistry	8	11	11	13	16
Dept. of Medicine*	5/64	17/74	29/87	26/91	53/114

Source: *UBC Calendars*. * regular/clinical (on-site teaching) appointments.
Note: Figures include the rank of Instructor and above, part-time, but not honorary appointments.

Part of the solution, Mike thought, to both the teaching and administrative problems lay in a thorough examination and reorganization of the Faculty of Medicine and his own department. Like students who protested against university policies and authoritarian administrative structures during the 1960s, Mike joined with other democratically-minded young professors at UBC and across Canada to critique their own institutions and to secure greater job security and administrative participation. (In the same spirit of advocacy, Mike, David Suzuki, Gordon Dixon, and others joined students in 1967 to protest the Vietnam War by endorsing and promoting a petition asking for an end to American bombing and a ban on arms sales to belligerents. Mike even joined a few peace marches until faced with the prospect of restricted border crossings.) UBC lagged behind some Canadian universities in adopting changes, and in 1969 Mike formally sided with the reformers when he joined the Personnel Committee of the UBC Faculty Association, an increasingly influential political vehicle.[24]

Mike's involvement with the Faculty Association began with a little self-interest. He had passed his first MRC review, but he knew that his salary was not part of the university's budget. He was paid from an external grant, "soft money" as it was called, leaving him without tenure and a little insecure about his career. Perhaps his experiences as a graduate student and with the Fisheries Research Board had left him careful not to assume anything about how others valued his work, especially when he depended on the goodwill of an autocratic administrator. In addition, Mike's reputation in 1969 was not yet broadly recognized outside of his specialty. A colleague in California noted that Mike had not received appropriate recognition for his work with Gordon Dixon although it was one of the best collaborations in North America. A historical overview of Canadian biochemistry published in 1976 included UBC professors Darrach, Dixon, Tener, Zbarsky, and Polglase (and FRB scientists Tarr and Idler) but failed to mention Mike. If administrators were not going to recognize and promote him on their own, Mike was willing to provide some encouragement.[25]

Early in 1969 Mike wrote to the President of the Faculty Association, Bill Webber, about his concerns. Webber, a graduate of UBC's medical school and a young Professor of Anatomy, was a moderate yet persistent participant in the politics of reform. Off campus he helped organize Boy Scouts and sports for children in his neighbourhood. Webber referred Mike's letter to the Personnel Committee — otherwise known as the grievance committee — which immediately invited the young biochemist to join and prepare a report on the problem. The School of Social Work had similar concerns about faculty members who were paid from special federal grants. Mike was easily persuaded to investigate, and he soon concluded that the university administration was reluctant to adopt new policies or practices that might change the situation. The problem was worst in the Faculty of Medicine where some fifty-eight faculty members (using 1967 figures) were paid with non-university funds and only eleven held tenure. Mike found that the policy on tenure for soft-funded positions was inconsistent across

Canadian universities, but none promised security to academic staff who were funded by external grants. Mike insisted that UBC needed a better policy to ensure that he and people in his situation would be treated like any other faculty member in regard to tenure, promotion, and benefits.[26]

UBC eventually improved its policy but never did provide a firm commitment to provide tenure protection to soft-money appointments: when the grant ended, the position was gone. Mike, however, was a special case, and in fact his department Head was working to ensure that this talented young researcher remained at UBC. Darrach approached the Dean of Medicine in October 1969 with a recommendation for promotion, noting that Smith had good research funding from the MRC (about $45,000 a year), supervised eight people, had already graduated one doctoral student, and had helped to develop and supervise an innovative teaching laboratory believed to be the first of its kind in Canada. According to Darrach, "Dr. Smith's energetic enthusiasm has been a driving force" in the department. A recommendation from Professor Charles Dekker, a molecular biologist at the University of California, Berkeley, affirmed that Smith's work was impressive. Dekker thought that Smith was eligible for a full professor rank at any American university, and hoped UBC would be successful in retaining his services.[27]

This was just the encouragement Darrach needed to convince his Dean that there was "no room for any action but an early promotion." In July, 1970, Mike was made a full professor and a year later he earned tenure, supported unanimously by department colleagues.[28] Presumably the granting of tenure meant that UBC administrators were willing to find a salary for Mike should the MRC not renew his Research Associate position. Mike was no doubt pleased with his personal victory, but he continued to serve on the Faculty Association's Personnel Committee until the spring of 1972. He was loyal to his colleagues and sympathetic to the broader problems they encountered, and he had many personal attributes appropriate to the committee: he was easy to approach, responsible, inspired trust, and cooperated well with others. The Committee called themselves "unsung heroes" for their discreet and effective handling of a wide variety of complex complaints.[29]

Although Mike had satisfactorily resolved his issue of promotion and tenure he was still unhappy with his broader work environment.[30] The perceived bias against his department and the authoritarian administrative structure of the Faculty of Medicine remained. Deans and department Heads were appointed by the Board of Governors, at this time effectively for life. Mike entered institutional politics more deeply when he was elected in 1970 to the Committee on Faculty Organization to survey opinion in the Faculty of Medicine and propose administrative changes. He may not have asked for the job, but he found it hard to say no to requests from his friends. Mike and other members of the Committee suggested a Faculty Executive with elected representation of department Heads, faculty members, and students. They also recommended balanced representation of basic science and clinical departments, a position no doubt Mike strongly supported. The reformers also proposed committees for nominations, curriculum, evaluation, and admission, all with elected faculty and student representation where appropriate. UBC Senate had endorsed the principle of student representation, and other faculties were also allowing students onto committees. Furthermore, Mike's Committee on Faculty Organization recommended faculty participation when awarding promotion and tenure, developing policy, and selecting Deans and Heads to serve limited terms. Above all, the Committee endorsed the position of the UBC Senate to have regular reviews of faculties and departments.[31] Here Mike and his colleagues were successful.

One of the first departments to be reviewed was Biochemistry since it had one of the longest serving Heads. Like many of his colleagues, Mike objected to the tradition of lifelong administrative appointments and he was dissatisfied with the Head of Biochemistry, Marvin Darrach. Mike may have wished to avoid confronting the person who had hired him and arranged his promotion, but like others he thought that teaching duties were unfairly distributed and that department leadership was wanting. Besides, the

American Medical Association had also recommended five-year terms for department Heads, a review mechanism, and a re-examination of teaching responsibilities. Darrach tried to stall the Biochemistry review until the appointment of a new Dean of Medicine, but others in the department saw no need to delay. An opportunity soon arose to avoid a messy confrontation when Darrach, who was due to retire in a few years, suffered a heart attack and took a sick leave. Sid Zbarsky served briefly as Acting Head before Jim Polglase accepted the role.[32]

The review of the Department of Biochemistry proceeded in the spring of 1973. The reviewers included three professors from UBC's Departments of Microbiology, Pharmacy, and Paediatrics, and their assessment was highly critical. The report highlighted funding and space problems, unfair distribution of duties, poor faculty and student consultation, lacklustre teaching and inadequate consideration of medical students, an overall "fair" research record, and discontented graduate students. Balancing this mediocrity were a few good teachers and distinguished researchers with international reputations. The strongest criticisms were directed at the department's leadership that was found to be autocratic, arbitrary, and responsible for low morale.[33] This was a far cry from 1954 when department members were seen as having "a highly progressive outlook and a stimulating spirit of experimentation."[34] The 1973 reviewers expressed hope that "the new chairman may help to rekindle certain fires that have gone out, but this may be asking too much."

Members of the department reviewed the report, criticizing the methods and timing of the evaluation and attributing problems to conditions beyond their control. Mike may well have been one of the few who agreed with the report, as he was certainly not the object of criticism and was a good friend of at least one of the reviewers. The department's graduate students also endorsed the report and identified as strong members of the department Smith, Tener, Beer, and Bragg (all recently appointed under MRC auspices); Gordon Dixon had recently left UBC. They identified four others as weaker department members although various circumstantial considerations applied. The students criticized two senior faculty as poor teachers with low professional standards and they suggested that laboratory space should be reallocated in favour of the more productive laboratories, such as those of Drs. Bragg and Smith.[35]

Mike and the reformers achieved some of their goals. The Faculty of Medicine instituted a review policy and began appointing Heads and Deans for limited terms and with faculty consultation. At the same time, the Faculty Association won the right to representation on the Board of Governors and later, in 1975, entered into a collective agreement with the administration. Attempts to certify the Association as a union failed when put to a vote; many in the Faculty of Medicine opposed unionization because of perceived conflicts with their roles as licensed physicians or independent professionals. In fact, most UBC professors favoured a "special plan" of collective bargaining instead of a union.[36] Changes in the university were aided by the 1972 election of British Columbia's left-leaning New Democratic Party that passed an unprecedented number of legislative bills in its short term as the government, including revisions to the University Act.

Mike was satisfied with the changes in his department and faculty, although various problems remained. He now frequently and vigorously expressed his opinions in the regular department meetings and wrote letters to express his views to administrators and anyone else he thought appropriate. With new faculty member Dennis Vance, Mike also coordinated the only committee with student representation, the graduate student liaison committee. All agreed that a new Head was needed who could energize and inspire the research and teaching agenda of the department. A few people thought that Mike would make a good department Head, but, although flattered, he declined all such invitations after the reforms.[37] The rising number of science students called for an increase in the number of faculty members and a revised, separate course for medical students. Laboratory space was still in short supply as the increasing complexity of biochemical research required larger laboratory teams, but unless additional resources were made available it would be difficult to find more room or attract new faculty or a Head from outside the department.[38]

Amidst all the institutional politics, Mike applied in the autumn of 1974 for another sabbatical leave. His real interest was science, not administration, and he wanted to explore another new area of his field. Fred Sanger's visit and subsequent correspondence about sequencing the genome of the phiX174 virus — the entire DNA of the virus — piqued Mike's curiosity.[39] Mike reasoned that it would be useful to be able to sequence genes — determine the exact order of nucleotides — identified and purified on his oligomer-cellulose columns. Sanger welcomed Mike to visit and work in his laboratory in Cambridge, England, if the necessary arrangements could be made. The Acting Head reminded the MRC of the significance of the destination ("because Mike was probably too modest"): Sanger's laboratory at Cambridge was "one of the most distinguished centers in the world for molecular biology and . . . open to only the best scientists." The MRC approved an eight-month leave with pay that included a living allowance, a children's allowance, a research allowance, and airfare for Mike, Helen, and their children. After some haggling and the use of holiday time, Mike was able to extend his stay to nearly a year.[40]

Mike was by then a well-regarded and well-liked member of the international community of biochemists and molecular biologists. He had established a productive laboratory with amicable colleagues and successfully encouraged administrative changes in his favour. But he had not yet distinguished himself with a scientific breakthrough. Perhaps part of the problem was the eclectic nature of his research, with studies to describe salmon sexual maturation, isolate natural hormones or unusual DNA, and synthesize nucleic acids. Mike's sabbatical in Cambridge redirected his scientific focus and provided the spark for the breakthrough that would earn him a place among the world's great molecular biologists.

4

SCIENTIFIC SUCCESS AT A PRICE

Mike's visit to England began an intense yet highly fruitful period of his professional life. His work in Sanger's laboratory put him at the forefront of research into the organization of genes and genomes and methods to sequence large DNA molecules. Mike returned from England as one of the world's leading molecular biologists with new ideas for exciting new research of his own. Back at UBC, he and his lab colleagues developed a method to manipulate DNA that would be recognized as a landmark in molecular biology. Mike also expanded his scientific pursuits into the commercial sphere, a result of public policy and personal ambition. Sadly, his single-minded devotion to science took a heavy toll on his personal life, and his years of greatest productivity were mixed with considerable personal distress.

Mike's period of intense research began when he and his family arrived in Cambridge late in September 1975. They rented a house nearby in Trumpington, a five-minute bicycle ride from the laboratory where Mike would work. Tom and Ian enrolled in a grammar school down the road where they were compelled to wear uniforms, and Wendy attended the local village school. Helen and the children spent many weekends exploring the countryside by bicycle and canal boat. Mike's brother Robin lived nearby and sometimes brought his family over to visit, joined by Molly and Rowland from Blackpool. Mike, however, completely immersed himself in laboratory work, spending up to eighteen hours a day at the bench. Sanger had a modest but inspiring presence, his students and postdoctoral fellows were congenial and cooperative, the science was exciting, and the late night beer flowed freely at the sports club. Although he was less dexterous than when he started his career, Mike was as hands-on as he could possibly be, explaining that even washing equipment gave him time to think. He talked science incessantly; he was in his element.

Mike's sabbatical goal was to learn how to sequence DNA, a new development that has since become relatively familiar. Sequencing permits detection of the precise order of the nucleotides (bases) in a DNA molecule. Knowing the sequence (as expressed in a series of letters, C, G, A, or T) helps to identify the DNA and to explain how the molecule determines the growth and development of the organism. Mike would isolate DNA fragments from a small virus called phiX174, bind each fragment (primer) to a region of an intact phiX174 viral genome, extend the primer by 150 to 200 nucleotides, and then follow a series of complex steps to identify the order of the bases that corresponded to the viral DNA. Other scientists in Sanger's research team would sequence their regions using the same complex "plus-minus" method. By combining the data from all the short sequences, ordered by virtue of overlapping sections within the fragments, the team would sequence the entire phiX174 genome of some 5,375 bases.

Mike had plenty of work ahead to test the new sequencing method, although preliminary studies promised success. The first step, purification of short DNA fragments to use as primers in the sequencing reactions, was time consuming. Strands of phiX174 DNA were cut into fragments using restriction enzymes that acted at very precise locations. The fragments were then separated from each other using a procedure similar to the electrophoresis that Mike had used during his Research Council days, but instead of sending an electric current through a saturated strip of paper to pull charged molecules along at different rates, the new technology used thin slabs of agarose the consistency of gelatine ("gels"). The mixture of fragments was placed at one end of the gel (anode), an electric current was applied, and the primers migrated through the gel towards the other end (cathode). The rate of migration is a function of the length of the molecule and the net molecular charge. After electrical separation, the discrete bands were collected — literally cut out of the gel — extracted, placed in a test tube, purified, and concentrated by precipitation.

For the next five months, Mike and Nigel Brown, a post-doctoral student, spent most of their time preparing primers and becoming familiar with Sanger's "plus-minus" enzymatic sequencing method.[1] They conducted some one hundred experiments between September 30, 1975, and March 16, 1976 — a rate of almost one a day. The experiments did not always have the effects Mike wanted. One early experiment "ran too fast" and Mike "decided [the chemical solution] wasn't hot enough." He abandoned that experiment. On October 6, he "lost the +A(OH^-) and +G(OH^-) during an alkali heating step — presumably the tubes were not sealed." Two days later, Mike recorded that "at this point, the whole experiment became a disaster — the gel leaked. I think I used the gel too soon as urea was leaking into slots even though I filled them as soon as buffer was added and also after I rinsed slots." In another experiment he noted "Lost the –C(Hae) and –T(Hae) samples at this stage. Also, in general panic mixed up the –A(OH^-) and –G(OH^-) sample. Set up at 18 mAmp 400v to see what happens to samples when spread all through the slot. Experiment didn't work."[2]

Many experiments did work, and of course one expects problems when developing complex new techniques. On another occasion, Brown noted that an experiment was working well but joked that the Z10 primer does not like being run backwards. Mike overlooked the advice, accidentally reversed the electrical flow in the gels, and ran the Z10 samples backwards for fifteen minutes, a mistake that would certainly compromise the experimental results. Sometimes, "the gel went wrong for some reason" but it was important to learn from mistakes. Mike even wrote down his questions and hypotheses, and reminded himself to follow his hunches: "this is a very 'iffy' idea but should check." On November 7, the results of his experiment were "all lousy" and were "thrown away . . . Must learn how to do the 2D electrohomo [electrophoresis followed by homochromatography] method reproducibly." As Mike became familiar with the procedures he made fewer mistakes.

Mike wrote research notes on December 22, 23, and 26, so presumably he took two days off for Christmas. New Year's Day 1976 was not a holiday, nor was it productive. He concluded in his latest experiment that "I wasn't thinking about what I was doing when setting up the expt – everything wrong." He and the family then took a two-week skiing trip to Italy, with laboratory work resuming upon his return. By the middle of March Mike had enough success in preparing the Hae primers to begin running preparative gels of AluI digested DNA.[3] He attached the primers to a template of intact phiX174 DNA, extended the primers into a longer oligomer using an enzyme, and added a radioactive tag.

Mike had to conduct two sets of chemical reactions to establish the sequence of his extended oligomers. The "plus" reactions degraded the oligomers, stopping at a known base. The "minus" reactions extended them, again stopping at a known base. After both sets of reactions, the contents of the test tubes were purified, separated using gel electrophoresis, and exposed on X-ray film to reveal visual patterns that identified the sequence of bases in the oligomers. Mike was generally pleased with his early results, although on May 24 he concluded that the experiment was not very good and on June 14 he noted, "DISCARD. Why is this wrong?" Each step of the complex process required careful attention to detail, and Mike had a tendency to do too many tasks at once or engage in distracting conversations with visitors. By September, however, he was again very pleased with results. He concluded that "the procedure now looks to be 100% OK. Try it out on HpaII and then on

HphI and lastly on AluI and this will complete the set of experiments needed for a paper on this subject."

The data produced by Mike and other members of Sanger's lab combined to yield the entire sequence of the phiX174 genome. It was very much a team effort. Essential to the success of this project was a new computer program devised by a colleague of Mike's brother who volunteered his time to write code using what was then an advanced method — cobol and punch cards — to organize the large amount of data generated by Sanger's sequencing team.[4] The long and intensive hours paid off with the elucidation of the exact order of the 5,375 nucleotides (labelled C, G, A, or T) making up the phiX174 DNA. It was the first ever enzymatic sequencing of a genome, an exciting accomplishment and a landmark development in molecular biology. Knowing the genome sequence of the virus meant knowing its most fundamental molecular properties, including the genetic instructions for all the proteins it made. This provided new insight into the basic nature of viruses but also promised to provide new insight into the genetic composition of any organism, including humans. Mike co-authored seven papers on aspects of the work, including his first major publication in the prestigious journal *Nature*.[5]

Mike and his associates also revealed another property of DNA, that it can be "read" up to three different ways, depending on where one starts to read. The genetic code is a three-letter code (corresponding to three nucleotides) and there are different combinations of three bases that can be read in each strand of DNA. An analogy might be to read the alphabet in groups of three, but starting at different points: abc, def, ghi . . . or bcd, efg, hij . . . or cde, fgh, ijk. . . . Mike helped discover the rare phenomenon known as "overlapping reading frames" that many viral genomes use to encode as much information as possible in a very small DNA molecule. The success of Mike's work completely justified his sabbatical in the eyes of his department Head at UBC, and Mike felt that the work was "the most fruitful and enjoyable scientific experience of my life."[6]

Meanwhile, Mike's laboratory at UBC continued to work on projects initiated earlier. Shirley Gillam was effectively in charge during Mike's absence although post-doctoral fellows generally knew what they had to do and would help in the supervision of graduate students. Mike later described Gillam as "one of the most capable researchers I have ever known" and kept in close contact with her by telephone and mail during his sabbatical. Laboratory staff soon became anxious about airmail letters, fearing another new set of instructions from their "boss" to alter their work or begin a new task. One of the lab technicians, Pat Jahnke, even visited Cambridge while on holiday where a very excited Mike provided her with a list of new equipment for the UBC lab. Staff from the Department of Biochemistry forwarded mail and took care of the administrative details needed to keep the laboratory operating smoothly.[7]

During his absence from UBC — or perhaps because of his absence — Mike was able to resolve a conflict that he had avoided for years. Mike set high standards for energy and dedication in his lab, but his full-time technician was not keeping the pace. Laboratory members had complained about the technician for some time but Mike just could not bring himself to dismiss a pleasant, sociable, and long-time employee. Now that he was half a world away, Mike tried to ease his technician out gently by denying him an increase in his weekly hours of work, presenting him with a small incentive to resign. When a visiting professor at UBC telephoned Mike in Cambridge to recommend a more direct approach, Mike finally wrote his technician a letter of dismissal.[8] Although he could be bold in his science or university politics, Mike always found it difficult to dismiss staff or make decisions that might upset people close to him, no matter how well justified.

While in Cambridge Mike met Clyde Hutchison III from the University of North Carolina. Hutchison was an authority on the biology of phiX174, the virus Mike and the others were sequencing. Over cups of tea the two discussed science, as usual, and Mike learned about the work in Hutchison's laboratory to insert DNA fragments from a mutant phiX174 virus into a host cell along with intact, single-stranded phiX174 DNA. The host cell, an *E. coli* bacterium, replicated to produce viral progeny some of which contained the mutation.[9] However, the method was limited to a few known mutations that were available and it was difficult to select for a specific recombinant.

Mike thought about these experiments and realized how his earlier work on synthesizing oligonucleotides might improve Hutchison's method. Perhaps a synthetic oligomer from his lab could be substituted for the mutant fragment. Mike was excited. He knew from earlier experiments that he could build a mismatching base into a complementary pair of oligomers. Could he create any mutation at any site within the viral genome? If yes, the result could be an efficient method to engineer heritable changes in genes. Anxious to test his idea, Mike called UBC and instructed his laboratory to make oligonucleotides that were identical to regions of phiX174 DNA except for one base in the middle.

By the time Mike had returned from Cambridge in the fall of 1976, his lab team had a good supply ready for testing. Gillam travelled from Vancouver to North Carolina to work in Hutchison's laboratory where Sandra Philips taught her the biology necessary to work with phiX174. Hutchison even visited UBC for a short while, helping to acquaint the Vancouver scientists with the procedures they would need for their new project. By the fall of 1977, members of Mike's laboratory were ready to begin the first experiments to create a mutation in the lysis gene in the phiX174 virus. Gillam bound the mismatched oligo to an intact phiX174 virus, extended it to form a complete complementary DNA molecule (except for the mismatch), and introduced it into *E. coli* for replication. In theory, if all went as planned, half of the progeny virus should have a mutated lysis gene and half should remain "wild-type." Unfortunately, the first experiment was disappointing but after analyzing the various steps for possible problems, the scientists revised the protocol and repeated the work. This time it was successful! They were ecstatic; 19 of 225 progeny virus particles isolated appeared to contain a mutation in the lysis gene. This was not quite the number predicted, but it was much higher than would occur simply by chance. The scientists used the second oligonucleotide to reverse the mutant virus back to wild-type.

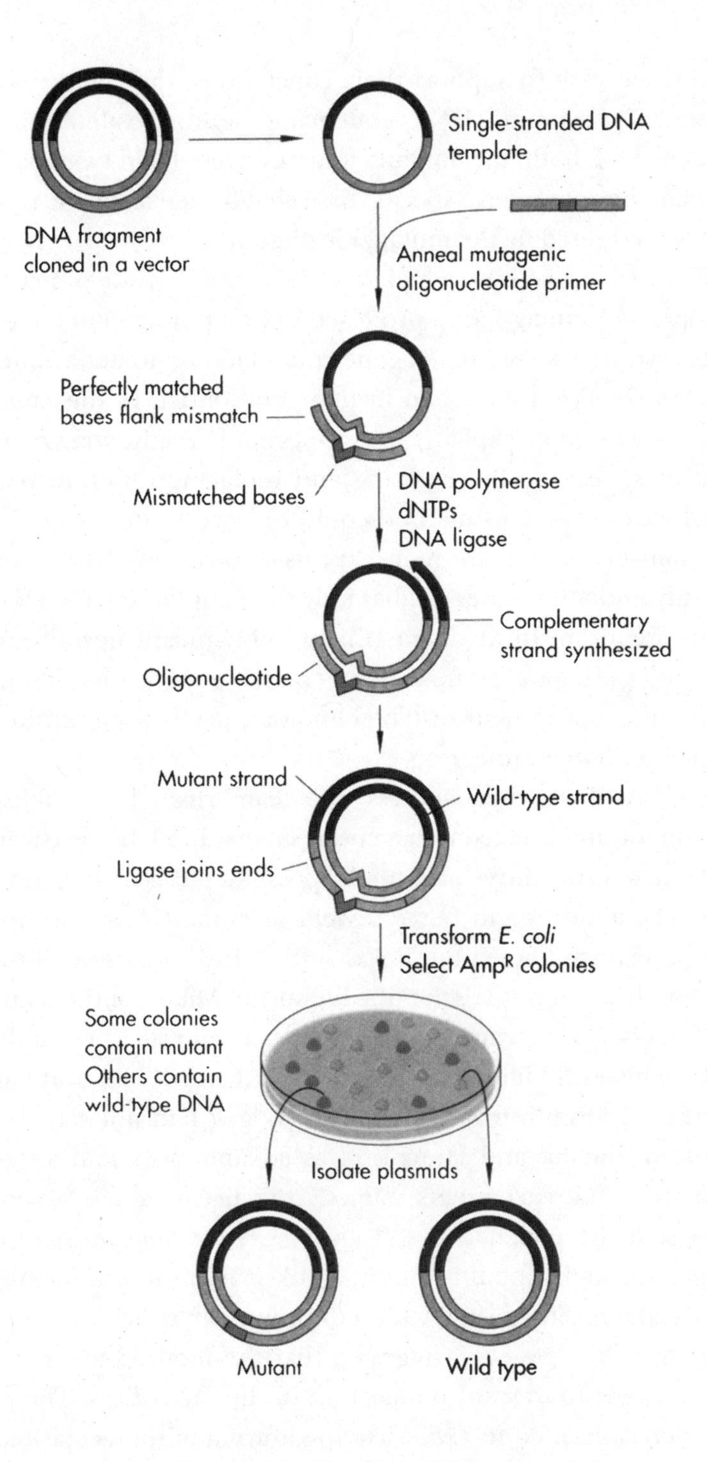

Steps involved in Site-directed Mutagenesis

Again, success! To support their conclusions, they used the cumbersome "plus-minus" DNA sequencing method to determine the sequence of both the mutant and reverted (wild-type) isolates. Indeed, they observed a specific base change at exactly the position that was targeted by the mutagenic oligomer.

The new procedure, which became known as site-directed (or site-specific) mutagenesis, promised to be a powerful new technology. A synthetically mutated gene could be used to determine how each nucleotide functioned in that gene, much in the same way (Mike would later explain) that a mechanic might analyze an engine by systematically removing and replacing different parts to see their effects. This method could be used to understand basic mechanisms related to gene expression and regulation, and to identify and alter genes deliberately to change the characteristics of an organism. In medicine, this possibly meant new diagnostic tests and perhaps even new treatments for genetic diseases. In the new and growing field of biotechnology, it was a significant step toward protein engineering.

Excited about their success, the team rushed to publish an account of site-directed mutagenesis since other labs were working on similar procedures and might possibly publish their findings first. The almost eight percent yield of mutant virus was not the fifty percent predicted but it was sufficiently high enough to warrant publication in the scientific literature. Mike lost the coin toss, so Hutchison's laboratory was listed first. Their article was authored by Hutchison, Phillips, and Edgell from Chapel Hill and Gillam, Jahnke, and Smith from Vancouver. (Edgell had not actually participated, but he and Hutchison as collaborators had an agreement to include each others' names, a practice no longer accepted.) They sent the paper to the *Proceedings of the National Academy of Sciences* through Gobind Khorana, Mike's mentor and member of the Academy. Surprisingly, the editorial peer reviews were disappointing. One reviewer suggested that the method was not sufficiently novel to warrant publication in the *Proceedings*. The paper was then submitted to *Cell*, a leading journal of molecular biology in which Mike had earlier published. The response was again disappointing: *Cell* also rejected the paper. The innovation Mike and his team submitted struck the editors as a technical detail too small to be of broad, general interest. These editors were unaware just how significant site-directed mutagenesis would become. Not to be deterred, the research team re-wrote and resubmitted the article to another leading publication, the *Journal of Biological Chemistry,* which published "Mutagenesis at a Specific Position in a DNA Sequence" in September 1978.[10]

Mike was more accustomed to having his research accepted immediately for publication, although rejections are common. Journal editors or peer reviewers often refuse papers because, in their views, the claims made are either too bold, too modest, or not appropriate to the journal. They may even disagree with the kind or amount of evidence presented, or even the form and style of the paper. The site-directed mutagenesis article went to prestigious, first rank journals that attracted many other submissions and thus were highly selective. Perhaps the reviewers did not understand how the oligomer based site-directed mutagenesis method could become a standard technology in molecular biology as few laboratories had access to chemically synthesized oligomers. Mike's group was in good company with their rejection. For example, Hans Krebs' work on the citric acid cycle was also initially rejected by the journal *Nature.* Krebs later won the Nobel Prize in Physiology or Medicine in 1953 for his work on the discovery of the citric acid cycle and was later knighted in recognition of his seminal work in the area of metabolism.[11]

While the journals were sorting out their priorities, work continued in Mike's laboratory to confirm that short synthetic oligomers could be used to make specific mutations in DNA. Subsequent experiments showed not only that one could change a specific base in the DNA, but one could insert or even delete bases. The efficiency of the procedure was improved to recover one hundred percent of the mutant progeny virus and methods were developed to identify mutants that had no physical properties that distinguished them from the wild-type. Mike had several more rejections from general, high prestige journals until a 1979 publication in

Gene brought his name and his lab's work greater scientific attention.[12]

Although developing site-directed mutagenesis occupied much of Mike's time, he was also busy introducing the sequencing technology that he learned during his sabbatical. Joining him at UBC in 1977 was his former doctoral student Caroline Astell, whose earlier studies on the thermal stability of complementary oligonucleotides had provided the foundation for site-directed mutagenesis. She returned after working at Rockefeller University, the University of Toronto, and the University of Calgary. Once back in Vancouver she helped arrange the equipment and establish the procedures for sequencing DNA using the new and improved enzymatic methods recently developed in Fred Sanger's lab as well as the chemical methods developed in Walter Gilbert's lab. Mike and Astell had advance knowledge of the "chain terminator method" that was proving to be relatively simple and highly effective. Mike's lab was soon Canada's foremost DNA sequencing centre.

During this period, Mike finished his research on marine organisms and was now concentrating exclusively on projects in molecular biology and how the genes within the DNA molecule act as reservoirs and transmitters of biological information. Experiments in the late 1970s were all related to developing site-directed mutagenesis, identifying and analyzing yeast genes, or sequencing DNA. His lab obtained the latest sophisticated equipment: a high performance liquid chromatography apparatus (to analyze the products of enzymatic synthesis of oligomers), a scintillation counter (for detecting radio-labeled molecules used in lab experiments), a high-speed centrifuge (for purification of biological materials including enzymes), and a computer terminal linked to the university mainframe. He brought the DNA software program that McCallum had written in Cambridge for installation on the UBC computer to help analyze the sequence of genes. In particular, Mike's lab was now dedicated to the new techniques of genetic engineering that had developed over the previous few years.[13]

All this new work at UBC was invigorating for Mike who, in his late-forties, still had tremendous energy. For a short while he was so excited that he quite literally ran about the lab singing, but this did not last long. His hands were a little more unsteady and now his eyesight was blurred to the extent that he had difficulty pipetting the very small amounts (e.g. 1/1000th of a millilitre) of solutions needed in his experiments. The long hours of standing were tiring. After a few months his work in the lab subsided, although his energy remained high. Instead, he spent his time organizing people and things, chatting on the telephone with scientific colleagues, securing grants, reading the scientific literature, reviewing progress in the lab, and, as usual, providing a plethora of ideas to members of his laboratory. His international scientific network was invaluable in keeping him at the forefront of developments. In return, he shared insights and materials from his own laboratory, even if he occasionally regretted promises to provide rival labs with supplies.[14]

Mike knew that he depended on good graduate students, postdoctoral fellows, and technicians to carry out the practical aspects of his work. When the Medical Research Council conducted its 1978 review, Mike once again received praise as a top-notch researcher, enthusiastic teacher, and valuable departmental resource. In his report, Mike emphasized that the studies were a group effort, and he thanked his research associates, visiting professors, collaborators, and other members of his research team. He told his old colleague Bill Webber, who was now Dean of Medicine, the same. His MRC support was renewed although his title was changed to "Career Investigator" for bureaucratic reasons. Following his 1983 report, it was decided that a site visit was not needed.[15]

Although Mike's laboratory was a busy, productive, and, for the scientists, exciting place, he knew that his department still suffered from some of its chronic problems. During his sabbatical in Cambridge, the British Columbia government decided to double UBC's class of eighty medical students to provide more locally trained physicians. UBC fast-tracked plans for new development before the deadline to spend Federal Health Resources Funds in 1980.[16] Although the medical school expansion provided new additions to the Copp Building and additional teaching and laboratory space for the Department of Biochemistry (by 1980 Mike had a new office downstairs nearer to his lab), there was no added provision for the ever-increasing number of science students. The Head of Biochemistry complained that they had a smaller budget per student than many science departments, a slightly decreased proportion of the Faculty of Medicine budget, a much higher teaching load, and inadequate staff support. Despite two new appointments and provision for a couple more, the total number of faculty members in the department had not risen adequately to meet teaching demand or the tremendous growth of research in biochemistry. Junior professors were tempted with offers from other universities. The department had been unable to recruit a new Head from outside the university because, it was believed, laboratory space was inadequate and salaries had not kept pace with the cost of living in Vancouver. Mike was not pleased and voiced his disapproval to those who would listen.[17]

Part of the problem originated outside the university with the economic recession of the mid-1970s that affected various government programs across the country. The federal funding scheme that for ten years matched provincial operating costs for higher education changed in 1977 to provide block transfers that provided less dedicated money for universities and demanded less accountability. The provincial government tended to allocate less money to higher education, and university funding in British Columbia dropped from nearly 6 percent of the provincial budget in the mid-1970s to about 3.5 percent by 1984.[18] Adding to what the Head of Biochemistry called "starvation amidst plenty" were reductions in Medical Research Council grants. Although Mike and many of his colleagues remained adequately supported by Canadian standards, MRC funding during the mid-1970s did not keep up with inflation as economic expansion in Canada slowed. The MRC had broadened its mandate with the result that grants were spread more thinly and favoured fashionable and less risky research. Some young scientists worried that their career prospects would suffer, and Mike also lamented the decreased vigour of the MRC program.[19]

Mike personally felt some of his department's continuing difficulties. He and other Medical Research Council Career Investigators (formerly MRC Research Associates) in the department worried that the MRC would not raise their salaries at a rate consistent with other faculty salaries. When UBC made up the difference (amounting to a 5.4 percent raise), Mike mentioned to his Dean in 1979 that this was the smallest increase he had received since joining the university, a net decrease of some five to six hundred dollars in the university's share of his salary from the previous year. His greatest concern, however, was that the university could not make up the deficit in his research grant that was overspent by nearly $20,000. He apologized to Dean Webber for griping, but expressed the hope "that an institute that always seems to be able to find money for footpaths and flowerbeds can help when there is a genuine need for funds in the one area which makes the institute a university, i.e. research."[20]

Although Mike complained about the changing financial environment and what he perceived as lacklustre institutional support for research, he was pleased to receive honours from the university. In 1977 UBC awarded him the Jacob Biely Faculty Research Prize, the university's top award for research conducted by a faculty member over the past few years. Prizes are won only with nomination and support from peers, and Mike had many colleagues who were eager to see his accomplishments recognized. Even Khorana (who had just received an honorary UBC degree) praised his former fellow's recent work as "breakthroughs." Mike was quick to acknowledge Shirley Gillam and his colleagues "who have to live with me 365 days a year."[21]

Despite the earlier UBC award, Mike had for some time been feeling frustrated that site-directed mutagenesis had not immediately gained recognition as an important and useful procedure and in 1980 he referred to himself as an "unknown from Vancouver." He knew this was not true, but deep down he still craved approval from his peers. Soon, however, site-directed mutagenesis would become better known and appreciated as he and his laboratory colleagues refined the method, published additional journal articles, and presented their work at conferences across the United

States and occasionally in Europe. Mark Zoller, a post-doctoral fellow who arrived in 1981, modified the procedure to use the bacterial virus M13 instead of phiX174. This provided a more general procedure to create mutations in foreign genes from any organism which had been cloned into the M13 virus genome. Mike's lab also had considerable respect from colleagues worldwide as a source of synthetic oligomers, although the new phosphotriester method of synthesizing oligonucleotides developed by Robert Letsinger and Marvin Caruthers (another Khorana post-doctoral fellow), would soon lead to an automated process.

In 1981, Mike finally began to receive the formal peer recognition that was so important to him when, after a nomination by Gordon Tener, he won the first Boehringer Mannheim Prize awarded by the Canadian Biochemical Society to recognize outstanding achievement by a Canadian molecular biologist. Mike received the award in Montreal, where he gave an acceptance speech. Well-wishers sent him notes of congratulations, delighted to see Mike's "initiative, humor and super good science" recognized. That same year he was elected to the Royal Society of Canada. Gordon Shrum, a long-time Fellow, congratulated Mike before the ballots had been counted.[22]

At the start of the 1980s, work was going well and Mike's professional reputation was climbing, but some aspects of his life had changed. Increased attendance at conferences meant that he was less available to members of his lab, and his collaboration with a colleague at the University of Washington on the genetics and biology of yeast required frequent trips to Seattle. (The collaboration led to the isolation of the gene for cytochrome *c,* a protein that plays a crucial role in providing energy for biochemical reactions.) Regular absences and a temporary drop in the number of successful, young students and post-doctoral fellows working under his supervision meant that lab parties with Mike had decreased. At home, parlour games with friends and colleagues were much less frequent after 1972 when Helen, with encouragement from her husband, resumed her work part-time as a technician in another UBC laboratory. They hosted few parties after their return from

Cambridge because Mike was so frequently at the lab or travelling rather than at home. In fact, he spent little time with his family even when his parents visited.

However, in 1981 Mike managed to plan another project, this time a commercial venture. The preceding year he had received several phone calls from American investors asking whether he would be interested in putting his DNA manipulation skills to commercial use. One caller from Wall Street offered him $5 million to start a biotechnology company. Mike was not very good at managing money — his lab was always in debt — and he hesitated, but the offers were tempting. He had known for years that the techniques to cut and recombine genes developed during the 1970s produced new biological products, and that fears about the safety of the new techniques were subsiding as utility and profitability of those products were growing. Small, entrepreneurial biotechnology companies were appearing across the United States to develop new pharmaceuticals, genetically modified foods, and other industrial biological products, attracting American university professors who were experiencing reductions in government research funding. Policy by the federal administration, particularly after 1979, further encouraged American researchers to pursue commercial activities. Ben Hall, Mike's collaborator at the University of Washington, also received offers from venture capitalists to fund biotechnology projects and he was prepared to act. University of Washington administrators were promoting cooperation with industry as a way to compensate for budget shortfalls and to support a faltering local economy, so Hall talked with Earl Davie, also of the University of Washington, and with Mike they decided to launch a biotechnology company in Seattle.[23]

Biotechnology was a risky business. Much of the science was unproven, intellectual property legislation was poorly developed, and investors were wary. Hall, an expert on yeast who had earlier consulted with industry, and Davie, an expert on proteins, found an investor to provide $500,000. With Mike, they founded a company called Zymos, later renamed ZymoGenetics, to contract with larger companies to find ways to manufacture various pharmaceuticals, including human insulin (produced in yeast) and proteins to coagulate blood. Mike, with his knowledge of nucleic acid chemistry, had vital information and skills to contribute. In fact, some of the techniques he and Hall had developed earlier were already in use by other biotechnology firms. Mike acted mainly as a scientific advisor for oligonucleotide synthesis and gene sequencing, helping to organize the laboratory equipment, hire and train staff, and solve technical problems. As usual, employees found him to be enthusiastic, supportive, concerned about their welfare, and genuinely pleasant.[24]

Mike probably had mixed motives for his involvement. He had always been interested in improving his income, a value he had initially learned from his entrepreneurial family. He was not wealthy, supported a family, worried about household expenses, and sent money home to his retired parents who had moved to Cambridge to be closer to their son Robin. But the prospect of profit was probably not the only lure. More important to Mike, as usual, was the opportunity to expand his science into a new and unexplored field, test the practicality of his research and, perhaps, make products of direct health benefit. Biotechnology in Canada was poorly developed, and for years Mike had considered other sources of funding to advance his work. Even with good grant support from the Medical Research Council, the money available to him was far less than that to large American laboratories.[25] As long as the commercial work was done on his own time — no more than half a day per week at UBC — he had the blessing of his Dean. In return, Mike had Zymos donate 25,000 shares to UBC in recognition of the university's role in supporting his earlier research, with the hope that one day the shares would be sold to support research.[26] Because it was a business, Zymos required non-disclosure and non-competition agreements. The company leased Mike a car for regular trips to Seattle (a white Mazda RX7 sports car with personalized license plates that read "Zymos") and paid him a consulting fee of $3,000 a month for several years.[27]

To prepare himself for this new project, Mike took another sabbatical from January to August 1982 to study at Yale University. His intention was to experiment with yeast, either cloning or introducing mutant genes into yeast cells. During his leave he wrote his Dean that he was having fun working in the lab fourteen to sixteen hours a day, seven days a week. "I seem to have infinite energy for doing experiments," he wrote, "and a great propensity for getting tired and frustrated by paperwork and committees." He felt that he was able to fulfil his job description only while on sabbatical.[28] However, instead of returning refreshed from his leave and ready for a new adventure in biotechnology, the fun he had enjoyed for the previous sixteen years came to a bitter end.

Mike had not taken his family with him to Yale. His children, Tom, Ian, and Wendy, were too old for the trip and domestic relations had been strained since the stay in Cambridge. Although Helen worked as a laboratory technician and Tom was beginning to study biochemistry at UBC, no one had the same intense dedication to science that Mike had. Ian and Wendy were not interested in research careers at all. Mike's friends were all scientists or academics in other fields, and his professional world dominated all his personal relationships. He did not share some of Helen's interests either, such as yoga and eastern philosophy that took her on several trips to India. (Mike once took a yoga class, but found it difficult to settle down and stop talking.) As his world and that of his family grew apart, Mike spent less time at home and became increasingly unhappy. As much as he hated confrontation and making unpopular decisions, he decided early in 1983 that he could not continue and moved to a small, simple apartment six blocks away. In his frustration, he denounced the institution of marriage as unsuitable for someone such as himself. He never did divorce.

The stress of the separation was almost unbearable and Mike sank into a serious depression that required medical treatment. Not since his problems as a doctoral student at Manchester had he felt so despondent, and he questioned the value of his life's work. He knew he had helped two other scientists win Nobel prizes — Khorana and Sanger, who won his second Nobel prize in 1980 for his new DNA sequencing techniques — but had not shared the recognition. Other Nobel near-winners in the past had felt similarly overlooked because the prestigious prize is awarded at most to three individuals, never an entire research team.[29] Mike was not much fun at work either. No one could predict whether he would be jovial or morose, and his anti-depressant medication made him speak a little too candidly with his friends about intimate matters. By the end of 1983 he felt tired. Disagreements over management structure and research priorities at ZymoGenetics had been stressful, and trips to Seattle for meetings, staff recruitment, and laboratory consultation added to Mike's fatigue, although these trips were decreasing as the company began to adopt routine practices. At age fifty-one, Mike felt a little older, less energetic, and less enthusiastic about his work. He resigned from his department's safety committee (he never did think it was a faculty responsibility) and proposed a reduction in the UBC portion of his salary to permit a decrease in his non-research work. He suggested that the savings should be re-directed to support a new instructor who, like several others in the department, was without secure funding.[30]

Just as he was getting over his depression Mike had one more upset. In January of 1984, two medical students, one a representative on the UBC Senate and one on the Curriculum Planning Committee of the Faculty of Medicine, lodged a formal complaint against Biochemistry 400. After a high rate of failures on the Christmas exam, the students alleged that the course lectures had not covered the appropriate material and the Department of Biochemistry had once again failed to serve medical students adequately. (The previous year, the course had a fifty percent failure rate.) The complainants suggested that the role of the department in the MD program ought to be reconsidered and perhaps terminated. The student complaints were consistent with a 1980 accreditation survey of the Faculty of Medicine (this time by the Canadian Medical Association) that gave the department a low mark for its educational provision, although the survey also blamed a high teaching load for some of the problems. However, the reviewers did approve of the revised Biochemistry 400 class that was now strictly for medical students and noted that further revisions were in progress under a new faculty member, Grant Mauk, who held both a PhD and an MD.[31]

Mike had been a principal lecturer for the course and was angry. He wrote letters to the Head of Medicine, the Head of Biochemistry, and the Dean of the Faculty of Medicine to defend the teaching of his colleagues and the effort spent in the past few years to revise the course. He questioned the maturity and responsibility of the student representatives, adding that they "caused considerable mental anguish to me and my colleagues." Suggesting once again that medical students did not appreciate the role of basic science in medicine, Mike thought that the complaints could damage the reputation of individuals and the entire department. He demanded an apology and urged disciplinary measures.[32] The Faculty of Medicine took the complaints seriously and eventually arranged for a curriculum review committee to investigate the courses in the MD program. The review of Biochemistry 400 released the following year recommended tutorials, student-centred learning activities, mid-term exams, better student orientation, and increased interdepartment consultation. However, the reviewers expressed strong support for the course in general and noted a tendency for students to inherit negative attitudes from previous cohorts. Mike, who had won a Distinguished Lecturer Award a few years earlier, received fairly good reviews from the ten students who completed evaluation forms in the spring of 1985, a year after the complaints. One of the faculty reviewers thought that Mike's lectures were clear, well-illustrated, and to the point.[33] Mike was largely exonerated, but it did not help to ease his frustrations as a teacher.

Mike eventually recovered from the emotional turmoil of 1983 and 1984, taking comfort in his growing scientific reputation as his work on site-directed mutagenesis proved to be increasingly useful. Various experiments in his lab showed, for example, how site-directed mutagenesis could be used to switch amino acids "on" or "off" to engineer desirable proteins or to alter the cellular mechanism that was believed to be responsible for cancer. Other experiments using site-directed mutagenesis showed that a particular

DNA structure in a herpes simplex virus gene was not responsible for regulating RNA synthesis, a discovery that eventually led to the first-ever characterization of a eukaryotic transcription factor. Mike collaborated extensively with Grant Mauk and Gary Brayer at UBC to explore protein structure and function, particularly how certain metal-binding proteins affected metabolism. Other projects using site-directed mutagenesis to study yeast genes helped solve the three-dimensional structure of wild-type yeast cytochrome *c*, which had defied solution for many years. Scientists worldwide began using Mike's method for their own work, sometimes visiting to learn from him directly, and increasingly citing papers from his lab.[34] The prestigious Cold Spring Harbour Laboratory in New York invited Mike and a few members of his laboratory to teach workshops on how to use site-directed mutagenesis. The new method was so popular that biological supply companies soon developed inexpensive "kits" that contained all the necessary reagents.

The success of these experiments and the growing importance of site-directed mutagenesis buoyed Mike's spirits and helped bring the fun back into the lab. Mike's laboratory (some dozen people now) moved in the early 1980s to a much larger room in a new wing on the third floor of the Copp Building, with his office directly across the hallway. He could thus visit the lab at various times during the day to talk about science with a fresh team of young and enthusiastic post-doctoral fellows and graduate students. Many of them had studied chemistry rather than biology, which Mike actually preferred. He sometimes joined them for picnic dinners on a nearby beach or for impromptu after-hours lab parties, and his habit of consuming ice cream quickly on hot summer days earned him the nickname "the vacuum." On other occasions he welcomed new laboratory members with a quiet homemade dinner in his apartment followed by a movie. Mike even began attending house parties again — usually arriving late — or joined his lab members at after-hours clubs in Vancouver for dancing; he always wore a wild costume at Halloween. A couple of times a year the whole lab went skiing and Mike always skied a few runs with the novices and the one renegade telemark (free-heel) enthusiast, Gary Pielak, who

▲ Strands of DNA, the genetic blueprint of life.
(COURTESY UBC ARCHIVES)

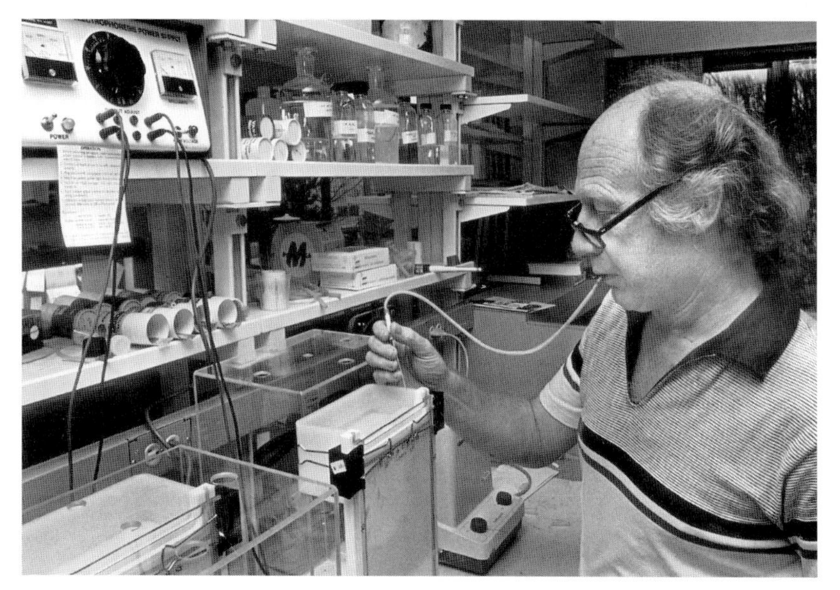

▲ Mike demonstrating lab procedures, 1981. Mouth pipetting is no longer an accepted laboratory practice.

(COURTESY BILL KEAY/ *THE VANCOUVER SUN*)

▲ Mike, in his classic short shorts, at Cold Spring Harbour, 1984, with a class in site-directed mutagenesis. Also from Mike's lab, in the back row, are Tom Atkinson, Mark Zoller and Gary Pielak.

(COURTESY UBC ARCHIVES)

▲ Mike in his UBC lab with Johnny Ngsee in the late-1980s.
(COURTESY UBC ARCHIVES)

▲ Mike at the official opening of PENCE, 1991.
(COURTESY UBC ARCHIVES)

▲ Mike at sea aboard the *Darwin Sound*, 1992, celebrating the winning of his Flavelle medal. Fellow crew members presented him with a canning jar lid in lieu of the actual medal. (COURTESY ELSIE WOLLASTON)

Mike with "Zymos," his second white sports car named for the biotechnology company he helped found. (COURTESY GLEN BAGLO/ *THE VANCOUVER SUN*)

Mike pauses in front of congratulatory cards and balloons to consider the significance of his Nobel Prize, 1993. (COURTESY UBC ARCHIVES)

▲ Mike, wearing his "How to Win a Nobel Prize" sweatshirt, with graduate student Mike Kalchman. (COURTESY UBC ARCHIVES)

▲ Mike celebrating his Nobel win with Helen (second from right) and their children Tom (left), Wendy, and Ian. (COURTESY HELEN SMITH)

▲ Mike accepts a one million dollar contribution to the Michael Smith Fund from B.C. Premier Michael Harcourt. (COURTESY UBC ARCHIVES)

▲ Mike with his guests in Sweden. Front: Mark Zoller, Shirley Gillam, Mike, Caroline Astell, Clyde Hutchison III. Back: Helen Smith, Karen Zoller, Robin Smith, Patricia Jahnke, Ian Smith, Lianne Comeer, Wendy Smith, Laura Wong, and Tom Smith. (COURTESY CAROLINE ASTELL)

▲ Mike accepts the Nobel Prize for Chemistry from King Carl XVI Gustaf of Sweden. (COURTESY CP/TOBBE GUSTAVSSON)

▲ Mike taking his place at the head table at the Nobel banquet. (COURTESY CAROLINE ASTELL)

claimed to have been inspired by Ernest Hemingway, one of Mike's favourite fiction writers.

As Mike grew older, he increasingly assumed the role of personal mentor rather than simply academic advisor, eager to inspire young scientists with a vision of first-rate science in a congenial and cooperative atmosphere. Research topics were diverse, but most of them used site-directed mutagenesis to investigate gene expression. Many of the young lab members had great fun together, sneaking into Mike's office on Friday afternoons to drink beer or playing snooker at the Faculty Club before dinner at a restaurant. Mike's lab still had its earlier shortcomings, however. As before, Mike rarely designed individual experiments to solve research problems, and now he was often out of town. Not every member of the lab thrived in an environment that placed great emphasis on self-reliance and leadership from the senior post-doctoral fellows, particularly when Mike's research suggestions did not work. A few graduate students left before completing their degrees.[35]

Mike's stature as a scientist was reaffirmed when he won several more awards. In 1984 he won the Gold Medal from the Science Council of British Columbia. Praise came not only for Mike's science, but also for the economic benefits that followed for the province's fledgling biotechnology industry. One supporter congratulated Mike and reminded him that he was considered "the brightest brain in B.C." and that his research was of great benefit to others. Another friend and fellow alpinist told Mike, "when you win the Nobel, remember your companions from ski-fit." Two years later, he was elected to the Royal Society (London), a particularly meaningful experience for Mike who felt honoured to sign the three-hundred-year-old book that held the names of Newton and Darwin.[36] Perhaps only Mike's mother, who attended the ceremony, was more proud. Sadly, Rowland had passed away the previous year.

In 1986 Mike won the Gairdner Foundation International Award for biochemistry, a major Canadian award that again recognized his work on site-directed mutagenesis. (Mike was on the Gairdner Award committee when nominated, but resigned temporarily while they deliberated the winner.) Mike told the local press that he would probably have to spend some of the $20,000 prize money on a fancy meal for members of his lab — a promise to which he was held! The wine flowed freely during lunch at Vancouver's prestigious Five Sails Restaurant, part of Canada Place built for Expo '86, and a good time was had by all of Mike's sixteen guests. Congratulations poured in from numerous well-wishers who were genuinely pleased to see Mike so honoured.[37] It was remarked that Gairdner Award winners often went on to win a Nobel prize and Mike was rumoured to be a top contender for that prestigious accolade. He was still not a flamboyant self-promoter, but he took care to have résumés handy at his invited lectures across North America and Europe. UBC's new President, David Strangway, proposed a dinner at his official residence to honour Mike, and soon after awarded him a Killam Research Prize. Mike was particularly pleased to win this award from his colleagues who knew him, as he said, "warts and all." He had his annoying traits, but colleagues in the academy where research achievement conferred considerable status were impressed with his work and found it difficult not to like him as well.[38]

As Mike's emotional health returned he entered what some friends called his "vanity period." He enjoyed driving his new, white sports car back and forth to Seattle and around town, although his preoccupation with talking science meant that safety-minded passengers often declined a second ride. He was, friends said, like a young boy with a new toy. Once, with a twinkle in his eye, he referred to the car as his "crumpet catcher" and fancied that women sometimes winked at him in traffic. His puckish sense of humour had returned. At other times, he could be seen riding his bicycle around campus wearing short shorts and sandals but without a shirt. For occasions that required more formality, friends tried to dress him up with new shirts and slacks, tight blue jeans, or a leather jacket. He was even more willing to wear a necktie when necessary. He began dating and found a congenial travelling companion for sailing trips and the occasional cross-country flight.[39]

As part of his new image, Mike attempted to improve a couple of his physical shortcomings. He started wearing hearing aids, which helped considerably when he had not misplaced or lost them. (He was once seen prowling a back alley looking for them after a theft from his car, hoping that the thief had pitched his hearing aids into a nearby garbage can. Several months later he found them in the pocket of a ski jacket.) Mike also had cosmetic surgery. His friends and colleagues had not seen him around campus for some time and rumour circulated that he was in a local hospital. Caroline Astell telephoned every hospital in Vancouver until she found a patient named Michael Smith and then took Shirley Gillam for a visit. He was delighted to have visitors but a little embarrassed to be seen with his mouth wired shut; his lower jaw had been surgically extended by some thirteen millimetres in an effort to correct the overbite that had caused him so much embarrassment as a schoolboy. In spite of his restriction, he still managed to talk continuously during the visit.

Mike even tried to improve relations with his family. When asked to propose a list of who might attend the celebratory dinner to honour his 1986 Gairdner Award, Mike asked if he could include his family. He wrote his Dean that "this sort of event is complicated by the fact that I live separately from Helen. We have a friendly relationship, and I would be quite comfortable if Helen, Tom, Ian, and Wendy were to attend. I am not sure how they feel. I know that whatever success I have had professionally owes an enormous amount to all of them." This was not the last time Mike sent invitations to his family, who were often present as his guest at special dinners or other events he attended.[40] Three years after moving out, Mike was intent on rebuilding some of the relationships with Helen and his children. He needed them for his own peace of mind, but he also felt a strong responsibility to support his family and to help whenever problems arose. The competitive drive demanded by the job during the late 1970s slowly gave way to a more patient and accepting attitude. In return, Mike's children, now in their twenties, began to understand why their father had so often been absent. Nonetheless, it was difficult for all parties to forget entirely their hurt feelings generated by separation.

By 1986, with his reputation higher than ever and his personal life a little more settled, Mike was considering his future. His role for nearly twenty years had essentially been one of a researcher running his lab, and, to a lesser extent, a teacher. This work for the most part had brought him considerable academic success and personal satisfaction. At UBC, his laboratory was pursuing interesting and important applications of site-directed mutagenesis while colleagues elsewhere were citing his work, collaborating in research projects, inviting him to teach seminars on his methods, and awarding him prestigious prizes. But as much as he enjoyed science and directing his laboratory he was spending little time "at the bench." Medical students still complained on occasion, leaving Mike weary. Despite his aversion to paperwork and committees, Mike began to think seriously about invitations to accept administrative jobs. He had never really wanted those sorts of responsibilities, but now he was concerned about promoting UBC's reputation as a centre of world-class research. It would be a matter of redirecting his energy to new developments at the university, and he again would impress his colleagues with his talents.

5

MAKING THINGS HAPPEN

It was not easy for Mike to exchange laboratory research for administrative duties. He loved science, hated paperwork, and avoided making decisions that might disappoint others. But he was now at the stage in his career when he was able to develop and oversee the implementation of important new projects. He served briefly on the UBC Senate in 1981 as the elected representative for the Faculty of Medicine, and in 1982 he supported and helped to launch the Centre for Molecular Genetics in the Faculty of Medicine as an interdisciplinary research and teaching unit. He also discussed plans to honour his mentor by establishing the H. Gobind Khorana Lectureship in Biochemistry, supported by Gordon Tener and others. Invitations to speak at conferences increased in the 1980s, and he joined several editorial boards, review boards, and national advisory committees.[1] Many of Mike's friends at the university were administrators, including the new Head of Microbiology, Bob Miller, a member of Mike's Friday afternoon social group and another former post-doctoral fellow of Khorana's. People at UBC and across Canada were looking to Mike for his views and guidance, especially in the new and growing field of biotechnology, and he increasingly wanted to provide advice. His sense of duty to colleagues — payback for all the years of fun, he said — and his dedication to science influenced his decision to enter administration, but certainly there were also circumstances within his institution that provided the impetus.

Mike was troubled by the deteriorating financial situation at UBC as the provincial government's fiscal restraint program of the early 1980s reduced funding for the university and public education more generally.[2] Opposition to government cuts steadily increased across campus as Faculty Association members supported the broader public protest. Matters came to a head in early 1985 when Pat McGeer, the minister responsible for universities, announced a freeze to the UBC grant and proposed direct government intervention in how part of the grant would be spent. UBC's Associate Vice-President Academic, Robert Smith, asked Deans of the different Faculties to defend their budgets and services; it looked as though programs would be cut, and faculty members fired. UBC President George Pedersen resigned in March 1985 to protest the province's low funding of the university.[3]

Mike had been worried about the fate of his department and wrote a strong note to his Dean, almost pleading for additional administrative support in the face of budget cuts. The untenured faculty members of the Department of Biochemistry had been anxious about losing their jobs since the hiring freeze of 1982. Several promising young professors were considering offers by other institutions, and one was actively looking for a more stable position elsewhere. An older professor was about to retire. With morale in the department low, Mike warned his Dean that if these people left and the positions were not held, the department might collapse.[4] Mike himself had a secure position with salary and research grant support from the Medical Research Council, but he had always been concerned about the welfare of others in his department, especially young scientists and their careers. He eventually wrote his local elected representatives to ask them for greater university funding, but he saw both the provincial Social Credit and federal Conservative governments currently in power as an unlikely source for renewed financial support.[5] Over the next few years, Mike would find alternative funding sources and be drawn into new government schemes to find financial "partners" to maintain research at his and other universities.

One way to support the department was to find research grants for new and different projects. They were "soft" funds, but they were better than nothing. Mike pursued that tactic when he co-signed a grant with the British Columbia Health Care Research Foundation in 1981 as "Principal Investigator" although he had no part in the research. Instead, the money supported his former student and now colleague, Caroline Astell, who became a faculty member in order to hold the grant. (Women were rare as faculty members in the Department of Biochemistry. Some thought the department was socially conservative, but over the years Mike quietly recommended hiring several women to tenure-track positions. Astell, however, was the first and for many years only female faculty member to hold a professorial rank.) The Centre for Molecular Genetics Mike helped to launch in 1982 was also intended to attract new funds from government granting agencies because of its potential medical and industrial benefit. It was hoped that new faculty could be hired as cross-appointments between the Centre and various departments, including Biochemistry. Mike also supported a Faculty of Medicine proposal to the MRC for a Biotechnology Training Centre Award to provide a research fellowship.[6]

Mike was not alone in looking for new research funds to support his department. The Dean of Medicine approached the Howard Hughes Medical Institute in 1985 with a proposal to establish a Lipid Research Institute that would keep Dennis Vance at UBC. The proposal was not successful and Vance left for the University of Alberta. Mike and the Dean of Medicine also lent their support to a proposal for research in evolutionary biology funded by a new body, the Canadian Institute for Advanced Research.[7] The CIAR had funds to support faculty in Canadian universities, and molecular biology was a qualifying area.

In early 1986 Mike was persuaded to become Director of the Centre for Molecular Genetics. It was having trouble — no funding had been secured and no outside Director could be recruited. Mike was a good choice to lead the project since his growing reputation was already drawing many high-calibre students and visiting professors to UBC, and the job would have relatively few administrative chores. Peter Larkin, Associate Vice-President, Research, favoured this appointment but took care to ensure that the MRC would continue to pay Mike's salary. Mike had been told that his involvement in the Centre was essential — or at least very helpful — in recruiting people and securing funding. Mike had always found it hard to say no when asked for help, and this time he knew how important the Centre could be in supporting his colleagues.[8]

Mike was, however, less enthusiastic about some aspects of the proposed appointment. Before he accepted the position, he told his Dean that the real problem was still lack of space for the Department of Biochemistry, a situation that made both senior and junior faculty members interested in job offers elsewhere. In 1983 the department had been placed sixth in priority on a list of Academic Building Needs by a UBC Senate committee. Mike even refused to participate in selecting a new Head of the department (now renamed the Department of Biochemistry and Molecular Biology) unless there was a solution to the space problem. In the meantime, Mike insisted, the laboratory vacated by Dennis Vance had to be kept for the new Head. (In the end, no outside Head was recruited and some of Vance's space was used to alleviate overcrowding in another lab.) Mike also wanted new space for the struggling Centre for Molecular Genetics. He would move his office only if there were firm plans for permanent new space, although he would begin recruiting earlier if temporary quarters existed. Above all, he declared, the Centre was not to be used to ease crowding in Biochemistry. His vision was to expand UBC's research capacities, not simply move people about: he was opposed to "putting old pots in new cupboards."[9]

Behind Mike's career shift were larger social and economic forces. As in other western industrial nations, Canada's post-war economic boom that had sustained his academic research program slowed in the mid-1970s and prompted restraint in government spending, including university research funding. As the slowdown continued, it brought in a new political mood that favoured decreased economic intervention by governments and increased reliance on the "free market." Legislation in countries such as Britain, Australia, the United States, and Canada encouraged universities after 1980 to compete more fiercely for students and government funds, to seek non-governmental "partners," and to adopt centralized, managerial decision-making procedures. Adoption of such policies in Canada lagged somewhat behind other countries and was less drastic, probably due to the decentralized nature of the university system. British Columbia's revitalized Social Credit government that returned to power in 1975 was likewise unwilling to fund public services generously, and after 1983 began a full-scale program of restraint in response to a world-wide recession. These broader movements accounted for some of the conditions Mike and his colleagues complained about in the late 1970s.[10]

At the same time, some Canadians voiced concerns about the ability of local industries to produce goods and services to compete in an international economic market. Mike was quite aware of the growing demand for value-added, high-tech commercial products. The Science Council of Canada recommended increased effort to couple scientific research with technological and economic development, as did the Natural Sciences and Engineering Research Council (NSERC), the federal government's new funding council. The Science Council of British Columbia worried that the Pacific province's long dependence on resource exploitation, especially timber, minerals, and fish, was vulnerable to unpredictable world demand. The Council proposed in 1982 to lessen this traditional economic dependence by encouraging new industries to adopt sophisticated technological procedures through closer cooperation among government, universities, and industry. Biotechnology was one of the new high-tech areas advocated both in

British Columbia and across Canada to compete with developments in other industrialized countries. Applied biotechnology research promised to improve food crops and forests, smelt mineral ore, treat sewage, pulp wood, improve detergents, and develop new drugs.[11] UBC was already a leader in such research, and Mike's work with ZymoGenetics had shown him that he could play a part. He and his colleagues anticipated biotechnology as the next big research wave, part of the "third industrial revolution" that included computers, robotics, and telecommunications.

Biotechnology, a new area of interdisciplinary research using genetic manipulation techniques, was not without its controversy. Most commentators agreed that the techniques created industrially useful products and that Canada lagged behind the United States and several European countries, but a few were hesitant to embrace the new procedures. Mike's old colleague, David Suzuki, warned of unanticipated and undesirable consequences, inadequate safety controls, and the spectre of human genotype manipulation. He was also worried that the ideal of curiosity-driven and impartial science would be compromised by the promise of profit; it had already infused science with greater secrecy and jealousy. Perhaps, Suzuki suggested, the best protection was first-class basic science funded by government in the public interest. He was not alone in his worries about the possible abuse of biotechnology for private profit, but by 1981 he agreed that some biotechnology products were valuable. Mike recognized these concerns (an ethicist was included on the Board of the Centre for Molecular Genetics), but was committed to the research and application of the new science.[12]

In the summer of 1986 Mike was doing what he could to improve financial matters at his university but the prospects for increased funding were bleak and frustrating. An offer to join the University of Manchester as the first Chair of Molecular Biology was tempting.[13] Still, he had family and close friends in Vancouver who were important to him, and he enjoyed his home near the mountains and ocean. He and various colleagues were doing internationally recognized research at UBC and could continue if only solutions to the financial crisis were found. He could not abandon the institution that had provided him with a home for twenty years without doing what he could to improve the situation.

Meanwhile, Mike's friend and colleague Bob Miller was working to provide an opportunity to expand biotechnology at the university. Miller was a determined scientist who had studied at the University of Pennsylvania before his post-doctoral fellowship with Khorana and subsequent appointment at UBC. He and Mike skied together and shared a friendly rivalry; Mike once affectionately referred to Miller as a "brash young faculty member."[14] Miller actively collaborated in a biotechnology project with several other UBC scientists and was not afraid of big projects. In fact, he had a big project in mind that went beyond what he could do as Head of Microbiology.

When Miller became Dean of Science in 1985 he immediately tried to persuade the provincial government to support biotechnology research and education. The government, however, was busy establishing the Biomedical Research Centre on the UBC campus, a new initiative funded through a partnership between the provincial government and a commercial pharmaceutical company, and paid little attention to Miller's proposal. Later that spring he persuaded UBC's Acting President to form an advisory committee on biotechnology, consisting of the Deans of Science, Medicine, Engineering, Agriculture, and Forestry. Mike was a natural addition to the committee. The Committee chose four potentially fruitful areas for research: fermentation (growing yeasts that would contain or expel useful proteins), process engineering (systems for large-scale production of biological products), plant science (to produce genetically modified plants), and molecular genetics (for pharmaceutical products).[15] Consistent with the political mood of the day, these areas all had commercial applications that promised to encourage economic growth.

UBC's new President, David Strangway, who joined the university in November of 1985, embraced the scheme. Formerly a professor of geophysics and an administrator at the University of Toronto, Strangway came to UBC promising to expand research and teaching in the face of reduced government support. He sought additional funds from industry, philanthropists, and alumni to make UBC "second to none" in research and teaching, and a vehicle for economic prosperity. He encouraged faculty to find new contracts and to develop "useful" technologies from their research that could be patented and licensed for profitable use, with royalties returning to the university. Strangway was remarkably successful in raising an endowment. At the same time, encouraged by other UBC scientists and administrators, he lobbied the government for money to support special projects, including biotechnology.[16]

In the spring of 1986, Miller put the finishing touches on a detailed proposal to the provincial government to acquire start-up funds under the Excellence in Education initiative. Unlike grant applications for basic research that described contributions to knowledge and the abilities of the researcher, subject to peer review, Miller's proposal outlined the wide-ranging industrial applications of biotechnology and its potential multi-billion dollar economic significance. He provided several examples of applied biotechnology identified in the National Biotechnology Strategy report, particularly in health, forestry, mining and smelting, agriculture, and waste treatment. All this high-tech activity depended on first-rate research in fundamental bioscience disciplines at a few key Canadian universities. Of course, Miller explained, UBC was one of those key universities. It already had a number of experts in the field whose work supported several local industries and Miller promised more such interaction. After Mike and other members of the Advisory Committee endorsed the proposal, Miller and Strangway visited Stan Hagen, Minister of Advanced Education, to convince him of the plan's merit. They were competing with the province's other universities, the University of Victoria and Simon Fraser University, and had to make a compelling case.[17] Miller knew of the University of Manchester's job offer to Mike and used it to strengthen his proposal, warning the government that without seed money for a new facility in biotechnology, British Columbia might lose one of its leading scientists and a potential Nobel laureate.

The proposal won approval and by June of 1986 UBC had been awarded $5.5 million over five years for a new project in biotechnology. This was only about half of what was requested, but the university was willing to redirect some of its budget into the new project and to find private sector funds as well.[18] Miller and his supporters thus launched the Biotechnology Laboratory, so named because it did not belong in any one faculty or department. Interdisciplinary centres that provided attractive working conditions for scientists were growing in appeal in the United States and other countries, particularly in "second-tier" universities that feared losing talent to larger institutions. The Biotechnology Laboratory at UBC joined the Centre for Integrated Computer Systems Research (CICSR) and the Advanced Materials and Process Engineering Laboratory (AMPEL) as one of the three provincial Centres of Excellence on the university campus.

Now the task was to make the Biotechnology Laboratory work. The Laboratory needed space, equipment and supplies, and a recruiting committee to hire ten first-rate people at junior ranks and cross-appoint them in participating departments. Whoever was chosen to coordinate the operation as Director would have to be of such a stature to attract these new faculty members. Mike had finally agreed to become Director of the Centre for Molecular Genetics but this project was to be subsumed by the new plan, making him a possible candidate as Director of the Biotechnology Laboratory although it was unclear whether he was interested and able to do the job. Despite Mike's limited administrative experience and hesitance to expand his formal circles of influence, the Advisory Committee on Biotechnology gambled and offered him the position late in 1986.[19]

It was not difficult to convince Mike to take the job — he certainly could not say no to his old friends. Bob Miller was eager to implement his scheme, and Bill Webber, as Dean of Medicine, was a diplomatic and congenial administrator who supported the new biotechnology initiative. Besides, as Director of the Biotechnology Laboratory Mike had an opportunity to strengthen his own department and research in biochemistry/molecular biology more generally. He would receive a salary supplement and administrative stipend totalling some $13,000 a year, a welcome bonus since consulting work for ZymoGenetics had declined over the past couple of years and the company had reduced his fees.[20] He might also have more influence to obtain permanent space to house the new facility, perhaps (Mike thought) to be named the H.G. Khorana Centre. Finally, Mike would retain his MRC salary (provided that administrative demands were low) and draw his UBC salary from the Biotechnology Laboratory budget. He could thus relinquish his teaching duties (thereby avoiding complaints from medical students) and leave his departmental income to salaries for sessional instructors.[21]

The Biotechnology Laboratory officially opened in January 1987, although its precise location in one of the biological sciences buildings on campus was not yet determined. As founding Director for a four-year term, Mike reported to the Academic Vice-President of the university through the Chairman of the President's Advisory Committee, who was in this case the Dean of Science. Mike's "boss" was effectively his old friend Bob Miller. Mike's central responsibility was to hire and support first-class scientists who were solving biological problems using genetic engineering techniques — cloning, site-directed mutagenesis, gene expression — and to encourage biotechnology activities on and off the campus.[22] Although industrial application lurked in the background, the Biotechnology Laboratory under Mike was to be a place of basic research.

Mike had to work quickly. He proposed a budget of nearly $2 million a year for the next four years, a level that rose slightly several years later despite a one-time cut of $700,000 in 1988–89. The first year's budget went immediately to refurbish space for laboratories and offices and to purchase communal equipment. In subsequent years, salaries replaced capital costs as the main expense although ongoing purchases of specialized equipment and supplies also required large sums.[23] According to the plan, Mike would hire ten new faculty members for the Biotechnology Laboratory, arrange for an academic cross-appointment in a participating department, and pay their salaries from the Biotechnology Laboratory budget. Much was riding on Mike's unproven organizational and administrative skill. Biotechnology research had become central to UBC's new, actively promoted image of itself as the source of knowledge to support competitive, high-technology industries in the province. Mike would share his success or failure with the university.[24]

Space was an immediate issue, and Mike insisted that the Biotechnology Laboratory should be as centrally located as possible. Initially, the Laboratory required seven faculty offices, instrument rooms, and research laboratories (each with benches and fume hoods, wash areas and space for instruments). It also needed rooms for administration, computers, a centrifuge, an autoclave, freezers, cold boxes, and a dark room. A greenhouse and teaching laboratory would complete the space requirements. To create space, the Dean of Medicine began relocating the Division of Medical Microbiology to the Vancouver General Hospital, site of several medical laboratories and a large number of faculty offices. This provided some 6,700 square feet of space for the Biotechnology Laboratory in the Wesbrook Building, near the centre of campus at University Boulevard and East Mall. Bob Miller took a careful look at the Biological Sciences Building across the road (he even measured the labs to see where space was being used efficiently) and decided that the Zoology and Botany Departments could be rearranged to provide 9,600 square feet of additional space. Mike dedicated his own laboratory space (1,900 square feet) as part of the Biotechnology Laboratory.[25] Douglas Kilburn, a Professor of Microbiology who also joined the Biotechnology Laboratory, dedicated his lab space to Laboratory projects.

Some of the established faculty members who were asked to move their laboratories were not enthusiastic and even jeopardized the schedule by blocking access to the design consultant. But renovations were not stalled for long. When the Academic Vice-President, Dan Birch, told Deans that the Biotechnology Laboratory was a priority, the project proceeded as planned despite ongoing haggling over space. The only major setback was a postponement of greenhouse and growth chamber facilities because of the 1988–89 budget cut.[26]

The success of the Biotechnology Laboratory depended above

all on the new faculty members. Mike struck a hiring committee that was small enough to be decisive but open to consultation with department Heads; he did not want to "parachute" unwanted candidates into departments. He fully supported a department's wish to hire a candidate directly, especially if the person came with external salary support. The Biotechnology Laboratory even helped departments hire new faculty by providing bridging salaries for up to two years or until department salary slots became available. But the Biotechnology Laboratory also needed new faculty of its own and within months the recruiting committee was scrutinizing candidates.[27] One of the novel attributes of the Biotechnology Laboratory was academic diversity. Mike's hiring committee interviewed molecular biologists, plant geneticists, and chemical and electrical engineers.

Mike wanted the best talent he could find, and he was choosy. He rejected those who were not specifically using molecular genetics techniques, and at times dug in his heels and delayed the selection process because he was dissatisfied. Even when he was satisfied, UBC's offer was not always accepted. The Biotechnology Laboratory had extra money, but it had limitations on salaries. Some potential faculty were offered annual salaries slightly higher than those offered by such departments as Biochemistry and Molecular Biology, but offers still remained below what other Canadian universities could provide. In the United States, corporate investment in biotechnology pumped tens of millions of dollars into individual universities, creating a "bidding war" for molecular biologists. Some new researchers were leaving UBC because they felt that the economic support for research in British Columbia was weak, a feeling no doubt exacerbated by ongoing legal battles between the Faculty Association and the university administration over salaries. Instead of high salaries, the Biotechnology Laboratory provided renovated facilities and fairly generous start-up funding of $150,000 per professor, about three times the amount offered by the Department of Biochemistry and Molecular Biology.[28] Although this was low by international standards, it was adequate to give Laboratory researchers a good start on their programs and hence their careers.

For Mike, the key to recruitment was to offer career satisfaction rather than simply a large salary. Scientists often say that money is not the prime motive for their work, and Mike had various ways to convince candidates to consider a position at UBC. President Strangway pledged to assist Mike by providing funds to sponsor candidate visits, help with moving expenses, and obtain housing on campus. Faculties were generally allocated a few units of university housing for new professors, but although the Biotechnology Laboratory was not a Faculty and Mike was not a Dean, he was able to secure this privilege. Mike promised candidates a reduced teaching load and on several occasions found university positions for spouses. He also promoted the quality of life in Vancouver by taking candidates to good local restaurants and driving them on scenic routes past local beaches, all paid for by generous recruiting funds in the Laboratory budget. Mike took his wining and dining seriously, and at times cancelled personal engagements to ensure that candidates were treated well. As a result, recruitment became a personal invitation; new faculty joined not a cold institution but a warm "family" with Mike at the head.[29]

Mike could be very persuasive when he gave his personal commitment to support the careers of new faculty members. He took great interest in candidates' research plans and promised complete personal and institutional cooperation. His enthusiasm was infectious and inspiring, and he demonstrated his honesty by disclosing various drawbacks about working at UBC. By April of 1988, Mike and his recruiting committee had appointed six new faculty members; two more followed a year later. The newcomers were promising young scholars, some previously known to Mike, who were given academic cross-appointments in Forest Sciences, Botany, Neurological Sciences, Microbiology, Medical Genetics, Electrical Engineering, and, of course, his own department, Biochemistry and Molecular Biology. Only two were women, one as a faculty member and one as the director of the advanced teaching laboratory. Mike had no qualms about hiring qualified women but the field remained male-dominated. True to his word, Mike took care of the administrative details that would help facilitate the new faculty's work and in the process he earned a reputation as a bit of a "mother hen."[30]

Little more than two years after its launch, the Biotechnology Laboratory was running at nearly full capacity thanks to Mike's efforts. The new professors he had helped to recruit lectured in their respective academic departments, served on various committees, and ran well-funded, collaborative research programs that engaged over one hundred staff, post-doctoral fellows, and graduate students. By late 1989 the teaching laboratory was being used for two directed studies courses until a formal course in plant molecular genetics could be approved, and two further laboratory courses were scheduled for the autumn of 1990. The seminar program with visiting scientists was also a great success, attracting researchers from both the university and industry.

The Biotechnology Laboratory quickly became a resource for the entire university, providing access to a linear confocal microscope and, for a small fee, supplying researchers across the university with bacterial cells, animal cells, genetically engineered proteins, and oligonucleotides produced in the automated synthesizer Mike had acquired for his laboratory. The synthesizer was expensive — Mike and eleven UBC faculty members paid $72,000 from their research grants (Mike was actually unable to contribute because his grant was already overspent) — but the "gene machine" was vital to keep pace with new demands for oligomers.[31] Using the new phosphotriester methods of synthesis, technicians could set the machine to make overnight what once took months. The oligonucleotide synthesizer subsequently became part of the Biotechnology Laboratory's Nucleic Acid–Protein Service Unit (NAPS) that also sequenced DNA and analyzed amino acids. To support expectations by UBC administration that allied departments would develop their own priorities to complement developments in biotechnology, Mike spent funds in other departments (including Biochemistry) on equipment and bridging salaries.[32] Perhaps Mike's only disappointment was that he was unable to fund transgenic mouse research and thus introduce this new field of genetics to UBC.[33]

By many accounts, Mike did an extremely good job organizing the Biotechnology Laboratory and encouraging biotechnology across the university and in the wider scientific community. He had help, of course. The Laboratory was supported by the university's central administration, several Deans, and the provincial government. When Bob Miller became the university's Associate Vice-President, Research, in 1990, Mike gave him credit for establishing and supporting the initiative.[34] Mike also had excellent office staff who kept him organized, within the budget, and on task. His administrator, Darlene Crowe, provided invaluable support for her boss whom she compared to the White Rabbit from *Alice's Adventures in Wonderland,* always busily rushing about, always late, but always with a cheery hello and a mischievous glint in his eye. (Sometimes he could be a little absent-minded. On one occasion, he could not find an expensive piece of equipment he thought he had purchased, so he ordered another one. Some months later the missing item was found under Mike's desk where it had been used as a footrest.)

The Laboratory was also well funded thanks to the Excellence in Education funds and a portion of UBC's general operating budget. An additional infusion came in 1988 as Mike and the Dean of Medicine were discussing the difficulties of recruiting to UBC. Dean Webber noted that the shares from Mike's biotechnology company, ZymoGenetics, that had been received five years earlier could now be sold and invested to provide an additional $6,500 a year. Like most start-up biotechnology companies, ZymoGenetics had never turned a profit but revenue had steadily increased and prospects for commercial products were good. Moreover, Novo Industri, a Danish pharmaceutical company and early contractor with the Seattle entrepreneurs, had offered to purchase the Seattle company. When UBC's share value jumped from $44,350 to $275,000 the university sold and put the proceeds into a Faculty of Medicine endowment fund for biomedical research, especially biotechnology, at the discretion of the Dean of Medicine, the Vice-President, Research, and the Director of the Biotechnology Laboratory. Over the years, the fund was used when necessary to keep Biotechnology Laboratory salaries competitive with those of other institutions.[35]

Of course, Mike gained personally when he sold his own shares in ZymoGenetics although he almost missed the opportunity. After a fretful morning pacing the floor and rummaging through filing cabinets he left the office in search of his stock certificate. When he returned a few hours later he was calmer. Mike had found the certificate stuffed in the back of a drawer at home, not in his safety deposit box as he expected. He was a major shareholder of common stock that was now worth millions of dollars, which explained his anxiety. (Still, he was annoyed that he had been paying for an empty safety deposit box all those years.) Friends wondered why he would keep such a valuable document in a drawer — they were never sure if it was a kitchen drawer or a sock drawer, or whether he kept his socks in a kitchen drawer — but they were pleased with his good fortune.[36]

For the first time in his life he was wealthy. Mike moved out of his Point Grey apartment and into a pleasant but modest condominium in the more colourful neighbourhood of Kitsilano. He bought himself a new bicycle for commuting, gifts for family and friends, and a large, gold bracelet with an image of a Haida bear carved by Clarence Mills. The bracelet seemed a little out of place with his casual wardrobe, but friends generally accepted it as an expression of Mike's individuality and new financial freedom. He began taking adventurous journeys to the Queen Charlotte Islands as a crewmember aboard the *Darwin Sound,* a seventy-one-foot ketch formerly owned by Graham Kerr, television's "Galloping Gourmet." Mike loved piloting the ship in rough waters, sharing his knowledge of marine life, and cooking in the ship's well-stocked galley. Fellow crewmembers thoroughly enjoyed Mike's spirit of adventure, good fellowship, and hospitality in the kitchen, never suspecting his eminence as a scientist. On other trips to the Charlottes, Mike stopped at the Langara Fishing Lodge where his daughter Wendy and later his son Ian worked.

He also bought a chalet at the nearby ski resort of Whistler, filled it with Ikea furniture, recreation equipment, and other supplies, and invited family, friends, and members of his laboratory, the Biotechnology Laboratory, and ZymoGenetics to use it free of charge. He told the scientists in Seattle that if it had not been for them he would not have the wealth, and the least he could do was to share some of it with them. He discovered good restaurants and fine wine, often treating his friends to an expensive meal. Despite his gifts to himself and others, Mike was careful with his new wealth. With the help of a financial advisor, he invested much of his money in fledgling British Columbian pharmaceutical companies and generally maintained his down-to-earth manner and casual lifestyle.[37]

Mike's personal generosity and the success of the Biotechnology Laboratory sent his reputation to a new high. Long-time colleagues were impressed with his previously hidden administrative abilities, and new recruits to the Laboratory were impressed with his dedication to their welfare. Their new jobs even came with subsidized ski weekends! Office staff were likewise impressed with Mike's honesty and dedication that more than compensated for his forthright manner and sometimes acerbic sense of humour. His office administrator was pleased when Mike defended her against an aggressive salesman. Another staff member was so impressed with Mike as a boss that she left him a poignant note when she changed jobs. She wrote, "As I see you conduct yourself with everyone (I mean everyone) I have learned that you treat each person regardless of rank with respect and honesty. This is something which I will try very hard to maintain."[38] Mike was generous and Biotechnology Laboratory staff over the years were touched with his small acts of kindness — one new dad, for example, quite unexpectedly received a stuffed toy for his daughter from his boss. Despite his administrative success, Mike still found it difficult to discipline or dismiss staff, sometimes waiting until he was out of town before telephoning anyone with bad news. By doing so, he hoped to return to the Laboratory well after the staff member had left, thus avoiding direct confrontation.

The Biotechnology Laboratory was only one of the new administrative projects Mike had in the late 1980s. Mike joined the advisory committee of the Evolutionary Biology Program of the Canadian Institute for Advanced Research, a network of Canadian scientists scattered across the country. Although the CIAR was dedicated to basic research for the sake of knowledge development, it was, like the Biotechnology Laboratory, investigating the science underlying a new technology and the new "knowledge economy."[39] Because most of the scientists in the program studied molecular phenomena, Mike was of great value in evaluating and shaping their work. Colleagues found him to be critical, compassionate, and humorous as he debated the goals of the Institute and how best to allocate funds. The CIAR scientists also worked to promote genomics, the sequencing and analysis of an organism's entire DNA, and particularly Canadian participation in the Human Genome Project.[40]

In January 1988, rumour circulated at UBC of a new federal government program to fund "Networks of Centres of Excellence." Mike was one of the first to be alerted and was soon drawn into a discussion with colleagues at UBC and across Canada about combining expertise in protein engineering through genetic manipulation.[41] Scientists at UBC, the University of Alberta, and the University of Toronto eventually proposed a project in protein engineering called the "Protein Engineering Network of Centres of Excellence," or PENCE. Mike played an important and personal role in drawing together participating scientists and in writing the proposal.

The national networks program was intended to combine the expertise of scientists working at universities, government institutions, and private laboratories. When it was first announced in February 1988, government bureaucrats attempted to direct the program to serve narrow industrial interests. Leaders of the scientific community scrambled to add peer-review to the process to ensure high standards of academic quality and thus make the program more acceptable to other scientists. Thus any network in the national program would ensure both "excellence" (basic research) and "relevance" (utility to industry). After lobbying by the scientific community, research quality became the most important criterion for judging proposals (fifty percent). Linkage or networking among universities, government, and industry constituted only twenty percent of the criteria, as did relevance to future industrial competitiveness via product development, dissemination, or opportunities for the private sector. Administrative and management competence of the proposed network accounted for the last ten percent of the evaluation criteria.[42] A dozen or more networks would be chosen according to the strength of the proposals, funded for a trial period of four years.

Despite encouragement to form industrial partners, participants in Mike's proposal for PENCE came mainly from three participating universities (British Columbia, Alberta, and Toronto) and two National Research Council laboratories. There simply were few people in private enterprises in Canada who wanted to or were able to join the project. Five companies initially joined the network, including biotechnology, pharmaceutical, and computer companies, but several were already affiliated with a participating university. The companies altogether promised that five of their own researchers would work on PENCE projects, compared to sixteen from the universities and five from the NRC labs. The university scientists noted that they would continue to do research that had no foreseeable commercial application and suggested that whatever product development might ensue could take fifteen to twenty years. In short, although Mike and his colleagues supported the notion of industrial relevance, they were not about to re-orient their research programs to short-term studies directed by industry or government. The proposal included a handful of other industry linkages — Kodak, for example, was interested in some of the protein research coming out of UBC — but these were small, transitory concerns. Other networks in the national program were also dominated by academic rather than commercial concerns because the scientists participated early in the government initiative and helped establish the rules.[43]

It took a year for the network proposals to be evaluated and selected. During that time, in late 1989, Mike enhanced the PENCE proposal by committing Biotechnology Laboratory resources to PENCE projects, thereby showing that the British Columbia government would be a participant in the network. Mike also used the Biotechnology Laboratory indirectly to increase the likelihood that UBC would obtain specialized new research instruments if his proposal succeeded. Upon learning that the national award might provide large sums for nuclear magnetic resonance equipment if his network had the expertise to use it, Mike offered Biotechnology Laboratory money for a new faculty member to study macromolecules using relevant techniques if the Departments of Chemistry and Biochemistry would eventually assume financial responsibility. Mike had rejected such a scheme two years earlier because he thought that the university resources were inadequate, but now he acted quickly. PENCE received funding, UBC received large grants for the specialized equipment (even though the University of Alberta was the "host" institution), and Lawrence McIntosh was hired as the NMR faculty member. In 1990, Mike became the scientific leader of PENCE, coordinating twenty-one scientists in three universities, two government laboratories, and six companies with a budget of $20 million over four years.[44]

Mike also supported an unorthodox scheme to find new office space for the national network programs at UBC. Other scientists at his university participated in different Networks of Centres of Excellence and one, the Canadian Genetic Diseases Network, was headquartered at UBC. Because there would be no funds for new buildings, Mike, Bob Miller, and other proponents of the Networks program asked Ron Woodward, the provincial Assistant Deputy Minister responsible, if UBC could borrow money for new construction and lease the new facilities back to themselves using Networks money. Woodward thought a minute and finally answered that such a plan would be acceptable. New offices were built above the UBC Bookstore thanks to the innovative funding arrangements approved by the province.

Mike's own laboratory in the Department of Biochemistry and Molecular Biology continued to do good work under his overall supervision, although his presence in the lab had decreased even further. Fortunately, he had an experienced crew of post-doctoral fellows, visiting professors, and technicians who could keep an eye on the doctoral students. Mike was apologetic but students and post-doctoral fellows were nonetheless frustrated by his absence. Mike had large grant support, more than anyone else in the Biotechnology Laboratory (over $224,000 from the MRC alone in 1989), and his lab published more than those of his new colleagues. Much of the published work was conducted by other lab members, but, as supervisor, Mike's name joined the list of others. He also participated in numerous advisory and editorial committees across the faculty.[45] When the MRC reviewed his laboratory in 1988, he again received stellar commendations for work using oligonucleotides as tools in molecular biology and on DNA sequencing, gene expression, and protein structure-function analysis. The new department Head described the work in Mike's lab as very high quality that set new standards. The MRC reviewer added that Mike's work was a "classical case of a highly academic and curiosity-oriented research pushing the frontiers of technological development in genetic engineering." The reviewer then noted that "it will not be a surprise if Dr. Smith is eventually awarded a Nobel Prize in medicine."[46]

Mike continued to receive awards, including his first honorary doctorate in 1988 from the University of Guelph. He began his address to convocation with earthy advice on how to give such a speech: it should be "like a woman's skirt or like a man's kilt for that matter . . . long enough to cover the subject but short enough to be interesting." He then thanked his laboratory colleagues and students for their contributions to his research success, and, recalling the scholarships that made his own schooling possible, condemned the policies that pushed Canadian students into debt. (A few years earlier, Mike had chastised Arnold School for having abandoned scholarships. Arnold's Headmaster defended his school by noting that the removal of the direct grant and the eleven-plus exam were political decisions beyond the school's control.) In his talk, Mike expressed the hope that science students would take a more active role in politics and use their scientific skills in non-polluting industries. Finally, he urged all students to retain a sense of humility.[47]

Also in 1988, Mike won the Award of Excellence from the Genetics Society of Canada. Congratulations again poured in. One newspaper article circulated the opinion that site-directed mutagenesis was the "intellectual bombshell that triggered protein engineering." Another well-wisher told Mike, "next step . . . Sweden." The following year, he won the G. Malcolm Brown Award from the Canadian Federation of Biological Societies, again for site-directed mutagenesis. As with other awards, these required nominations and Mike's colleagues at UBC were happy to put his name forward.[48]

Mike himself remained dedicated to first-class, basic research which he promoted as the foundation for any subsequent application. He had also told his University of Guelph audience in 1988 that "all great discoveries are made by mistake, in my experience." He agreed that PENCE could be "the kind of enhanced research and technology transfer project the government has in mind" but that it would take time.[49] He continued to publish in the standard academic journals and refused to release research findings to industry before it had been published. The new federal copyright legislation discussed after 1987 roused Mike to lobby for free discussion of research results within the scientific community via published reports. He encouraged his graduate students to follow their curiosity or the natural evolution of their research.[50]

Although his main interest was in basic science intended to expand knowledge, Mike was learning to exploit the resources increasingly available from industrial or commercial sources. Some of these new funds came indirectly through the national funding councils that were also looking for non-governmental sources of money. The MRC was forming industrial partnerships in the mid-1980s while NSERC, which superseded the independent National Research Council as Canada's primary scientific funding council, was administered by a federal ministry. Government policy now influenced funding more directly, and policy after 1987 dictated that federal grants to NSERC were now dependent on industry contributions. As much as Mike preferred "curiosity-driven" research, he knew that most funding had to come from these councils and he encouraged Biotechnology Laboratory scientists to apply for whatever funding they could obtain, including NSERC Strategic Grants that had explicit industrial objectives. A few scientists in the Biotechnology Laboratory eventually won Strategic Grants or held contracts with government and industry.[51] Over the years, the Biotechnology Laboratory generated commercial activity as it gained a reputation not only for internationally recognized research but also for "an enviable record of spinoff company creation, patenting, and invention disclosures."[52]

Mike's willingness to find industrial funding for basic scientific research went even further. As a member of the Biotechnology Sector Committee of the Science Council of British Columbia he supported the application of biotechnology to fish biology, food crops and livestock, human infectious and hereditary disease, fermentation and process engineering, and the genetic engineering of forests. As well, he recognized a role for "imaginative" university-industry relationships, "imaginative" investment, intellectual property law, and health, safety, and environmental regulation. Mike accepted the necessity of cooperating with industry, but insisted that university scientists should retain their prerogative to conduct independent, basic research.

The growing popularity of applied biotechnology took Mike beyond British Columbia. Singapore was interested in sending scientists to Canadian laboratories, and the MRC supported the initiative. Mike was approached in 1990 to facilitate a liaison with the Institute of Molecular and Cell Biology at the National University of Singapore, and he was "happy to initiate and foster this project." Mike facilitated scientific exchanges for many years as coordinated by the UBC President's Office. (Much later, the Singapore contact admitted that without Mike he probably would not have collaborated with UBC.) Like other UBC initiatives at the time, such collaboration was not simply academic but intended to develop intellectual property and commercial opportunities for Asia-Pacific and American markets.[53]

Mike welcomed the quick rise of biotechnology research at his university and its appeal to commercial funding bodies, believing that this was important and useful research that had to be done whether governments supported it or not. American social scientists observing similar changes in their universities were less enthusiastic. They warned that overly close ties between universities and industry could compromise the intellectual autonomy and the ideological credibility of their institutions. Canadian academics — especially in the humanities and social sciences where budget cuts were not so easily replaced by funds from outside the university — soon joined the criticism, arguing that industrial contracts, spinoff companies, patents, investment in outside companies, and even consulting could unduly encourage or discourage lines of research, influence the interpretation of results, and suppress free publication of data and conclusions. What would the public think of "objective" scientific opinion if university proponents of biotechnology — Mike included — had to consider economic benefit for themselves, their institutions, and clients? Some even questioned whether university intellectual property agreements really were good for the general economy, and others warned that biotechnology was too poorly understood to be safely applied at all.[54] With an emphasis on "excellent" scientific research, what would become of "average" (but still high quality) science and teaching?

Mike's activities also fit with the broader trend toward centralized university administration that saw the "democratic" mood of the 1970s subside in the face of economic anxiety. Mike had been appointed as Director of the Biotechnology Laboratory by a Deans' committee, which became the advisory committee, and he reported to the Vice-President Academic. Decision-making by small, centralized committees rather than through consultative and democratic processes was reinforced when the government dedicated a line-item in the UBC grant to the Biotechnology Laboratory, which, as Mike had hoped, retained sole control of salaries. Administrators at UBC were not particularly enthusiastic about line-items since the university had little say in how the money would be spent. The Biotechnology Laboratory Director also had considerable personal influence in the promotion and tenure of its scientists.[55] Such arrangements facilitated useful freedom from bureaucratic restraints but raised the concern that faculty might lose their prerogative to help select and evaluate their colleagues. A few UBC professors also wondered whether cross-appointments might weaken department support. The federal Networks of Centres of Excellence prompted similar concerns since scientists were accountable to the directorate in Ottawa although dependent on their host institutions for space and salaries. Despite these centralized administrative structures that gave him considerable authority, Mike by his nature was very careful to consult before making hiring and tenure decisions.

Although Mike was re-appointed Director of the Biotechnology Laboratory in 1990, he still maintained ties with the Department of Biochemistry and Molecular Biology through his primary academic appointment. He also held a prominent place in their brochures and advertising — nearly half the accolades cited for the department in 1990 were Mike's, and six were his alone. Although he generally identified with the Biotechnology Laboratory and, because of time constraints, did not often attend meetings in his old department, he still made his views known in his characteristically forthright manner. In early 1991, after reading the minutes of a department meeting he had missed, he chided members for what he felt was a reluctance to embrace the interdisciplinary trends of the day as represented by the Biotechnology Laboratory and other similar initiatives. "If the Biochemistry Department persists in its present attitude," he warned, "it will be conspicuous by its non-participation. King Canute would have been wise to learn how to swim rather than to try to order the tide to recede at his whim. . . ." Some colleagues were amused and some were annoyed, feeling that Mike had over-reacted to a technical motion to place restrictions on who could legitimately be called an Adjunct Professor in the department. At the very least, they thought, he could have attended the meeting![56]

In the spring of 1991 Mike accepted a new mission as the Interim Scientific Director of the Biomedical Research Centre, an interdisciplinary research unit at UBC that was having administrative problems. Once again, Mike was asked to take on this temporary role by his old friend, Bob Miller, the Associate Vice-President, Research. Once again, Mike could not say no to the request, even though the Biotechnology Laboratory advisory committee was unenthusiastic about this appointment.[57]

The Biomedical Research Centre was a unique laboratory that had entered a period of controversy. The BRC had operated on the UBC campus since 1987, and although its scientists had cross-appointments in various academic departments, funding came from the independent Terry Fox Medical Research Foundation.[58] The Foundation — no relation to the Terry Fox Cancer Run or Terry Fox Cancer Laboratory — was launched in 1981 by the provincial government to raise funds in support of research, particularly in biotechnology and pharmaceuticals. Burroughs-Wellcome, the Canadian arm of the British pharmaceutical giant, joined with the Fox Foundation as an equal partner in 1986 to construct the BRC building on the university campus, and to conduct research primarily into the efficacy of the anti-cancer drug Interferon. Mike had supported the BRC project in the early 1980s before Burroughs-Wellcome became involved and before he put his energies into the Biotechnology Laboratory.

The Fox Foundation had been having administrative problems for several years and Mike knew it. A 1989 review by Price Waterhouse accountants noted that the Foundation was a highly complex mixture of agreements between provincial, federal, and corporate funders. The report warned that the usual administrative and financial checks and balances were not in place and the Foundation's chairman possibly had conflicts of interest. Millions had been spent on consulting fees with little result. Moreover, the commercial value of Interferon — one of the drugs tested by BRC scientists and which cost the Foundation $20 million — plummeted as it proved to be less efficacious in combating cancer than originally thought. Perhaps the biggest problem, according to the Price Waterhouse report, was that the Foundation was attempting to be an entrepreneurial business venture rather than a research and development organization, and lacked the expertise and investment to be successful. A crisis developed early in 1991 after Burroughs-Wellcome announced that it would not renew its five-year contract and the Foundation board terminated the contract of the BRC Director. Politicians accused Burroughs-Wellcome of unfairly profiting from public investment, the BRC Director protested his dismissal as unfair, and BRC scientists feared a collapse of the laboratory despite a recent positive review of their work. When news of the internal politics reached the public in the summer of 1991, a few months before a provincial election, the issue became steeped in partisan accusations of mismanagement and deceit.[59]

Mike thus had several political problems if he were to help preserve the Biomedical Research Centre. He had to present an operating plan and budget that would demonstrate the viability of the laboratory and attract renewed government funding. This would require a public campaign to counter allegations that the BRC had been corrupt.[60] In October 1991, a second report circulated that also criticized the Fox Foundation for mismanagement but downplayed conflict of interest charges on the grounds that trustees had not been deceived. Again, the problem seemed to be insufficient expertise in the pharmaceutical industry and inadequate funding. Fox Trustees countered with claims that the scientists at the BRC had done very good work, and that government meddling had played a significant role in the problems. Mike criticized parts of the second report as irrelevant, inaccurate, or inadequately explained. Uppermost in his mind was the preservation of the Centre, and this seemed likely after Burroughs-Wellcome transferred its shares to UBC and a newly elected provincial government promised to provide adequate funding.[61]

The other political problem Mike faced was internal. Many of the BRC scientists still supported their former Director, whose research was integral to the operation of the laboratory. They thought his dismissal was conducted improperly and was probably unwarranted in the first place. Furthermore, they suspected that Bob Miller, with Mike's help, was trying to absorb the Biomedical Research Centre into the university and particularly the Biotechnology Laboratory, and they were opposed to such an idea.[62] Mike spent a lot of energy convincing them that he was not conquering new territory for a biotechnology empire. (He was probably telling the truth. The following year, Mike rejected a suggestion by UBC to combine the BRC with the Biotechnology Laboratory because such a move would undermine his credibility. In addition, he felt unsuited to continue as the BRC Director.) The former Director, who carried on with his research as a faculty member in the Department of Medicine, had not wanted to leave the job and in Mike's view actively worked to discredit those trying to save the Centre. Mike felt "blackballed" by most of the BRC staff and more than a little hurt, especially when he heard the rumour that no one would attend the Christmas party if he did.[63]

In the end, UBC assumed control of the Biomedical Research Centre and the provincial government closed the Fox Foundation in March 1992. Mike was willing to cooperate with Michael Harcourt's newly elected New Democratic Party government and visited politicians and bureaucrats in Victoria several times to present a case in favour of the BRC. The government decided to provide direct funding similar to that of the Biotechnology Laboratory, but at a reduced level which fortunately was still adequate to ensure good research.[64] Although an Acting Director had been appointed and procedures for selecting a new, permanent Director were in place, matters were not entirely settled. Over the next few years, BRC scientists continued to criticize the administration of the Centre, especially hiring procedures for the Director's position. A few resigned from the BRC to protest, while others noted with alarm that Burroughs-Wellcome had retained the right of first refusal to products developed at the Centre. Despite the politics, the BRC scientists were able to continue with well-regarded work and the original Director, whose scientific reputation was never in question, was reappointed several years later.[65]

Mike, however, was finished with the BRC in the spring of 1992. He was pleased with his success in preventing its collapse, but disappointed that the mess had arisen in the first place. The job left him physically and emotionally drained — he was usually liked by

▲ Mike, with Elizabeth Raines on his left, shows Prime Minister Jean Chrétien his Nobel Prize, 1993. (COURTESY PETER BREGG/ *MACLEAN'S* MAGAZINE)

▲ Mike unfolds the poster commemorating his prize-winning Nobel discovery. Henry Friesen, President of the Medical Research Council, on the far right. (COURTESY UBC ARCHIVES)

▲ Mike taking the helm aboard the *Darwin Sound* in 1994.
(COURTESY AL AND IRENE WHITNEY)

▲ Mike in his University of Manchester PhD gown with Diana Crookall (Bragg), Biochemistry Dept. Administrator, preparing for the 1994 convocation ceremonies at UBC. (COURTESY RICHARD BARTON)

▲ Mike with a big one! Langara Fishing Lodge, Queen Charlotte Islands, 1995. (COURTESY WENDY SMITH)

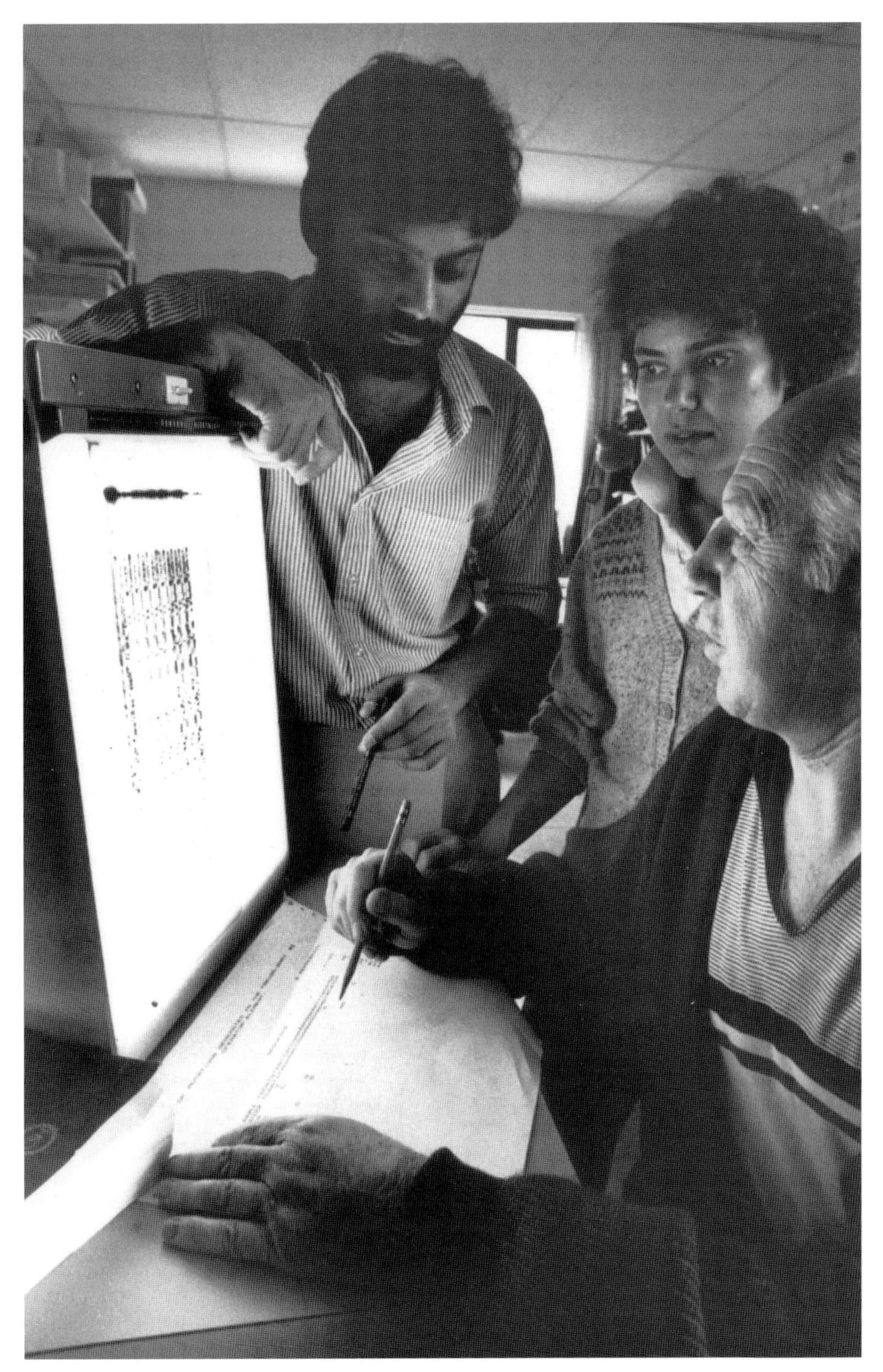

▲ Mike reading DNA sequences with Sarbjit Ner and Jeanette Beatty.
(COURTESY UBC REPORTS)

▲ One last party in Mike's UBC laboratory, 1996, before its closure. Left to right: Lianglu Wan, Ricky Chan, Bjorn Steffensen, Louis Lefebvre, unidentified, Steve Rafferty, Rob Cutler, Guy Guillemette, Dave Goodin, Hailun Tang, Robert Maurus, Susan Porter, Janet Yee, Chris Overall. (COURTESY *UBC REPORTS*)

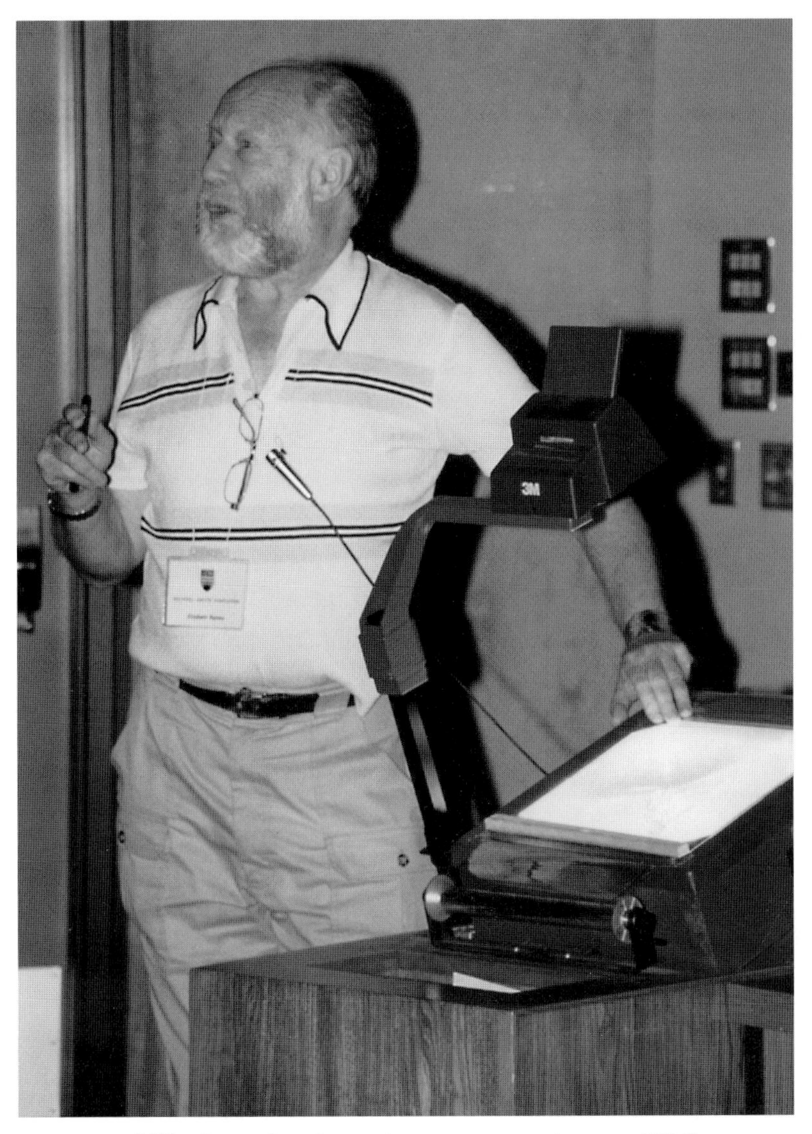

▲ Mike in action lecturing on genomics at a UBC symposium, 1997. (COURTESY RICHARD BARTON)

▲ Mike's sixty-fifth birthday dinner with (left to right) George Mackie, Elizabeth Raines, Mike, David Ward, and Rich Roberts. (COURTESY RICHARD BARTON)

▲ Mike with Marco Marra, Head of Sequencing, and Steven Jones, Head of Bioinformatics, at the Genome Sequence Centre, February 2000, on the occasion of Marco's receiving the Outstanding Alumni Award from Simon Fraser University. (COURTESY MARCO MARRA)

colleagues. He made last-minute arrangements for a holiday in May aboard the *Darwin Sound,* this time for the trans-Atlantic journey he had long dreamed about. He enjoyed the smooth waters from Florida to Bermuda but thrived in the storm near the Azores, and had fun at the improvised ceremony held on board to honour his latest award, the Flavelle Medal of the Royal Society of Canada. He even escaped the telephone calls from UBC that had too often interrupted his holidays in the past few years although he departed at the Azores to return home. During the trip, however, he learned that his travelling companion of the past six years, who was also on board, did not share his intention to further their relationship. Mike easily talked about science or politics, but found it difficult to discuss his emotions. Having finally and rather suddenly revealed his desire for a more committed partnership, he was stunned to realize that his personal plans would not proceed as he hoped. Although the parting of ways was civil and the friendship continued, Mike was unhappy.

When he returned home he felt so dejected that a friend insisted he have a party to celebrate his sixtieth birthday. Mike's daughter Wendy helped to organize the event that became just the right sort of festive occasion to boost his spirits. Mike invited everyone he could think of — including people from the Biomedical Research Centre — and some one hundred and fifty guests arrived between three in the afternoon and midnight, consuming ninety bottles of wine and a great deal of food. As had happened so often in the past, good friends and good cheer improved Mike's mood considerably.

Now he was better able to enjoy the success of his recent work. Mike had earlier thanked government staff and the Minister of Advanced Education for their commitment to interdisciplinary biomedical research and promised to share a celebratory beer. They now returned the compliment, thanking him for his untiring efforts to make the Biomedical Research Centre into a "silk purse" and for having the vision, patience, understanding, and enthusiasm that made it work despite the risk to his personal health, research, reputation, and free time. Bill Webber, newly appointed in the President's Office, thanked Mike for being as "ideal a faculty colleague as anyone could possibly wish to have." Webber praised Mike for his leadership in the Biotechnology Laboratory, especially in recruiting faculty, and for rescuing the Biomedical Research Centre from probable disaster. Webber added that this was all done "with determination — but in a manner which has enhanced the personal respect in which you are held by all who work with you." Mike thanked Webber for the kind words, adding that he did the job simply to support interdisciplinary biomedical research on campus and because he had been asked by a friend.[66]

Mike's agreement to work on behalf of the Biomedical Research Centre was mixed with another more secretive motive to raise funds to support research at UBC. He was learning how to use his money philanthropically and his contract with the BRC carried a condition that $100,000 be donated to UBC to help establish the H.G. Khorana Chair in Biological Chemistry. In fact, this donation was in lieu of his salary as the interim Director. Mike had already donated $85,000 to support his project ($70,000 raised through the sale of property he had purchased in the mid-1980s), and he added an additional $15,000 with the understanding that the money would be matched by the university. UBC launched a fund-raising campaign for the Chair, but Mike's gift remained anonymous so that he would not unduly influence his colleagues.[67] Endowed Chairs were not a major part of UBC's tradition, but their numbers were increasing.

For the rest of 1992 Mike returned to his usual work. The Biotechnology Laboratory was still growing, with a new faculty member having been recruited and $380,000 transferred to the President's Office for the greenhouses. The new Dean of Science, Barry McBride, reported to the UBC administration as Chair of the Biotechnology Advisory Committee that the Laboratory was running well and that research funding exceeded operating costs by fifty percent. Teaching was proceeding as planned and the first doctoral defence of a student working in the Laboratory was expected that year. McBride was another old friend of Mike's who tempered ambition for his university with diplomacy, but Mike insisted that the Laboratory must be reviewed rigorously and with critical honesty.[68] PENCE also required a considerable amount of Mike's time to write or edit proposals and grant applications, collect and present performance data, and meet with scientists, government officials, and industry representatives. Mike also felt obligated to visit the participating laboratories across Canada. The Network was a large, interdisciplinary enterprise of nearly two hundred people that included such research areas as physical chemistry, crystallography, and nuclear magnetic resonance imaging, areas outside Mike's personal expertise. He had much to learn.

That summer, Mike celebrated "landmark" birthdays of several friends including Gobind Khorana, who flew in from Boston with his wife for the weekend. They all stayed at Mike's new seaside home at Bliss Landing, north of Powell River on the Sunshine Coast some 150 kilometres northwest of Vancouver. The celebration of age was followed by a frightening reminder when Mike learned in August that he had a malignant melanoma on his leg. It was removed at an early stage but the skin graft did not take well and took some time to heal. He then underwent another surgery for a hernia which upset his plans for an early season skiing. But he could look forward to a ski trip over Christmas following his review of PENCE laboratories across the country.

In January 1993, plans for Phase II of the Biotechnology Laboratory began. Mike had always wanted a new, permanent laboratory complex; now it appeared that he would finally have a new building. A budget, set at just under $20 million, had the approval in principle of UBC administrators who were evidently pleased to hear that the Biotechnology Laboratory was well managed fiscally and administered effectively. A report early in 1993 noted that six faculty were to be evaluated in the coming year for promotion and tenure now that a procedure was in place. Departments were also conducting the peer and student teaching evaluations that were, after 1990, required by the UBC Senate. Biotechnology Laboratory faculty were considered "truly an outstanding group of scientists . . . performing at a very high scientific level." In addition, outreach activities continued to be popular with faculty and staff across UBC.[69] New facilities were all that remained to complete Mike's vision for the Biotechnology Laboratory.

In the meantime, there were still some administrative bugs of concern. Faculty members who held research grants or contracts traditionally had a small percentage deducted to cover operating costs in their department, and these funds were increasingly important when operating budgets declined. The Faculty of Applied Science was willing to split overhead charges on private contracts held by Biotechnology Laboratory cross-appointments, but Mike objected to levies on the federal grants when the work was conducted in the Biotechnology Laboratory. He even wrote to the Dean of Forestry to protest "taxing" the grant of a Biotechnology Laboratory member.[70] For Mike, this was evidence of a pressing need to formalize the ad hoc agreements that had hitherto guided the Laboratory.

Mike also wanted to establish a firm basis for faculty hiring, previously done through the Faculty of Science. He could see no need for departmental approval of grant-funded (non-faculty) appointments and wanted to clarify policy on sabbatical leaves, particularly who would pay for replacement lecturers. Since new department Heads were asking for higher teaching loads, a firm decision had to be made on teaching responsibilities for Biotechnology Laboratory faculty. Mike wanted more recognition in UBC's administrative networks so that he would receive all relevant information, and he wanted more exposure for the Biotechnology Laboratory in university publications. He also suggested a formal mission statement: to foster teaching of biotechnology, encourage basic and applied research in biotechnology, support biotechnology-related research across campus, and to enhance collaboration with biotechnology companies, especially in British Columbia.[71]

PENCE also required some reworking. The federal government had extended the program in December 1992 but cut its funding and re-emphasized the necessity of industrial collaboration. As it became apparent that the privileged place of basic research in the original PENCE agreement was in jeopardy, Mike spoke out against the renewed government emphasis on "targeted" research and promoted instead the view that society was better off when scientists were left alone to forage for new information that others could put to use later. Mike did not mind the commercial application of science (he eventually obtained a patent from his work to improve the oxygen-carrying capacity of myoglobin in blood) but he did not want university research to be directed by business leaders or politicians.[72]

Although the decrease in government funding for scientific research disturbed Mike, in other ways 1993 was shaping up to be a good year. The Khorana Symposium he helped to organize in May was a big success, attended by guests and old friends from across North America, Europe, and Asia.[73] During the symposium, Mike seemed to be especially spirited, and friends soon understood why when he introduced them to Elizabeth Raines, the woman with whom he was in love. He first met Elizabeth, a realtor on Vancouver's fashionable west side, when she helped him purchase his Kitsilano condominium and his Whistler chalet in 1988. For several months before the Khorana Symposium she had declined Mike's invitations to dinner, wondering what the two of them might have in common. When Mike suggested they attend a symphony concert he struck upon a shared interest, and soon they were discovering that they both enjoyed classical music, theatre, fine dining, skiing, boating, fishing, hiking, and even science. After their date to hear the Vancouver Symphony Orchestra, Elizabeth insisted on a second one to hear astrophysicist Steven Hawking who was on a speaking tour. After more than a decade of separation Mike was ready for a committed relationship and the two discussed plans for marriage.

Adding to Mike's good mood in the fall of 1993 was an invitation to attend his daughter's birthday party on October 12. He wined and dined at a downtown restaurant longer than he expected, arriving home to his apartment late. Sleepy from the martinis earlier in the evening, he removed his hearing aids, slipped into bed, and fell into a deep sleep, never thinking that he would awake as a celebrity.

6

AMBASSADOR FOR SCIENCE

Mike's alarm usually rang at 6:00 a.m., but he had overslept. At ten to seven on the morning of October 13, 1993, he awoke and turned on the radio hoping to hear whether the Toronto Blue Jays had won their baseball game. Instead, he heard news that changed his life: he was the co-winner of the 1993 Nobel Prize for Chemistry for his work developing site-directed mutagenesis. When the Royal Swedish Academy of Sciences had called earlier that morning Mike had missed the call because he was not wearing his hearing aids. Receiving no reply, the Academy mistakenly telephoned a very surprised Vancouver physician of the same name. But now the telephone started ringing again, first with confirmation from the Academy (which had verified Mike's whereabouts) and then with enthusiastic congratulations from well-wishers. Excited, Mike contacted Elizabeth and his children to tell them the news before setting off for his lab where others would be waiting to help him celebrate. By 9:30 a.m. UBC's Public Affairs Office had already logged fifty-six calls for information, and Mike's office fax machine had received a stack of congratulatory notes five centimetres high from around the world.[1]

Mike had long thought he might win this award. He had worked with several other Nobel laureates and his own work was proving to be just as significant. Respected colleagues world-wide agreed that Mike deserved a Nobel and had nominated him previously without success. Suddenly that Wednesday morning he felt tremendous excitement finally to have grasped the elusive Nobel. Although he often appeared modest when describing his work, Mike was in fact very proud of his achievements and now there was no question that he and his collaborators had made a substantial contribution to science. He also realized that the status conferred by the Nobel Prize would be useful to promote science in Canada, but for the time being he had to adapt to his new role as a celebrity.

When Mike arrived in his lab on the morning of October 13, 1993, friends and colleagues were already celebrating with bottles of champagne, waiting to shake his hand as he grinned ear to ear. His estranged wife, Helen, who worked in a lab down the hallway, also came to offer her congratulations. Television and newspaper reporters asked about his work and his response to the Prize. They were amused by his wardrobe — polyester slacks, well-worn polo shirt, and black socks poking out from under his Birkenstock sandals — and delighted to relate the story that he had been sleeping stark naked that morning. When asked whether he belonged to Mensa, the high I.Q. society, Mike answered that he did not and probably had a low I.Q. anyway — he did not consider himself to be particularly smart. His modesty, wit, and ready acknowledgement of laboratory colleagues and the country that supported his work made him a popular celebrity, and over the next few days he was the subject of numerous stories in newspapers, on television, and on the radio.

Overnight Mike became well-known, with reporters appearing in his laboratory looking for dramatic photographs while fans and well-wishers jammed the telephone line with their congratulations. Prime Minister Kim Campbell was unable to reach Mike's lab (she faxed her congratulations later), unlike a young woman who offered to do anything to meet Professor Smith. People on the streets, in pubs, and at stores recognized Mike and said hello, and on one occasion lined up to watch him at a local carwash. Another person asked him to record a song to support a social cause — Mike declined. He loved all the attention brought by the stories circulating in newspapers and on radio shows, but he did what he could to retain some humility; he asked his administrator, Darlene Crowe, to give him a good swift kick if his head started to swell. Winning the Nobel Prize overcame all the anxiety and self-doubt Mike had ever had, providing him with the highest of honours from his peers and complete acceptance by strangers. Needless to say, Mike's work was interrupted for the next few weeks; he was warned to be careful lest interruptions continue, but he found it difficult to say no to requests for his time.[2]

Of course, UBC was particularly proud of "its" Nobel scientist. Colleagues, administrators, students, and staff sent their congratulations. He was made UBC's fifth "University Professor," which meant that he could devote his time to research and teaching across the university and to matters of science and research policy as he saw fit.[3] Another UBC professor and administrator, Peter Larkin, composed an ode to Mike that was read at the B.C. Science and Engineering Awards dinner a week after the announcement:

> He says he has a low I.Q.
> But I suspect that isn't true,
> Perhaps the case is just that he,
> Is not the same as you and me.
>
> He doesn't have the kind of genes
> Of presidents, or heads or deans.
> His are those that live in sleuths
> Who go about unearthing truths.
>
> So starting soon, why not today,
> Let's take apart his DNA.
> Let's find the locus of the gene
> That makes his science lean and mean.

If perchance we find the locus
For his science hocus-pocus
We can clone all shapes and sizes
And win lots of Nobel Prizes.

Mike had not expected to be the guest of honour at the dinner and sincerely thanked the audience for the moving tribute.[4] Those who knew Mike were happy not simply because he had won, but because they thought the award could not have gone to a nicer person.

Mike had won other prestigious awards, but nothing that compared with the cachet of a Nobel Prize. Part of the award's allure came from the large monetary prize — Mike won nearly $500,000 (U.S. currency) — and part from a tradition of publicity that began in 1897 with fascination about the will of industrialist Alfred Nobel who had left his entire fortune to the prizes. The literature and peace prizes early roused national rivalries, and the science prizes garnered particular public interest after 1903 when the French press sensationalized the wretched living and working conditions of Pierre and Marie Curie, the winners in physics that year. Popular fascination with elite scientists (and the powerful technologies they initiated) became part of the Nobel legacy. Although the names of many winning scientists never became household knowledge, the Nobel Prize, whether in physics, chemistry, medicine or physiology, literature, peace, or economics, became popularly regarded as the highest honour in those fields.[5]

Such esteemed awards are never without their politics and the Nobel history is peppered with allegations of awards inappropriately granted or withheld. Much important work is never acknowledged by a Nobel Prize. Although the deliberations of the Nobel committee are kept secret for many years after the decision, Mike had a few advantages. All Nobel candidates require a nomination and the Nobel committee invites them from various reputable sources, including the heads of academic institutions and other laureates. Colleagues at UBC were especially eager to nominate one of their own. In fact, this was Mike's third nomination by Gordon Tener, and in 1991 both Phil Bragg and Harold Copp, two highly respected UBC scientists, submitted nominations. (A British scientist was believed also to have nominated Mike.)[6] Perhaps other laureates working in the fields of chemistry or molecular biology such as Gobind Khorana, Fred Sanger, Arthur Kornberg, Paul Berg, Michael Bishop, James Watson, or Canadians John Polanyi and Gerhard Herzberg — who all knew Mike personally — had provided support. At the very least, Mike was already a well regarded member of the international scientific elite.

Mike had a few other advantages as well, particularly the field in which he did his research. On thirty-three occasions between 1933 and 1993, Nobel prizes for medicine/physiology or chemistry went to DNA-related research. Mike shared his award with Kary Mullis, a maverick American scientist who had developed the polymerase chain reaction that allowed for the rapid and fairly simple replication of large quantities of DNA segments. The recipients of the Nobel Prize in Medicine or Physiology that year, Phillip Sharp and Richard Roberts (whom Mike had worked with previously), were themselves geneticists. Mike's academic pedigree may have been an advantage as well, since he had a Nobel mentor, Gobind Khorana, who had worked under Nobel laureate Alexander Todd. Being from England and working near the United States put Mike in a favourable environment, since British, American, and German scientists have won far more Nobels than those from other countries. His preference for laboratory work may also have given Mike an additional advantage, as experimenters have won many more Nobels than theorists, and the rise of commercial biotechnology may have increased the perceived value of site-directed mutagenesis. Finally, Mike lived long enough to be recognized. Prizes are often awarded many years after the ground-breaking work, but not posthumously.

Yet he also had a disadvantage. Although Mike frequently stated that world-class research could be done at UBC, the truth was that he had not done his work at an elite institution. Leading institutions like Cambridge, Stanford, or even IBM's Zurich laboratory are more likely to provide the environment that will keep a scientist at the leading edge of new discoveries, but Mike, perhaps through his extraordinary network of colleagues, had obviously been enough in touch with the international scientific community to make his unique contribution. Mike knew from personal experience that policy to support basic research in his home province and country had its shortcomings, but he now had strong evidence that UBC was, at least in some areas, a world-class institution.

Perhaps the most controversial aspect of a Nobel Prize is establishing propriety and priority. Was Mike truly the "discoverer" of site-directed mutagenesis and had he done it first? These are extremely important questions in an already competitive field. Several people had been involved in developing the technique from the beginning when it was discussed over cups of tea in Cambridge, and several names were on the first and subsequent papers. Exactly how the Nobel committee decided this matter may not be known for a long time if ever, but decisions generally follow the lead of the international scientific elite. Yet Mike knew he had not done the work alone, and although he was not able to share the prize (and certainly not about to decline!) he publicly recognized the colleagues whom he felt played important roles in developing the technique.[7] He even asked them to attend the Nobel ceremonies as his guests, paying for their transportation and accommodation. Mike invited former UBC lab colleagues Caroline Astell, Shirley Gillam, Patricia Jahnke, and Mark Zoller (who had returned to the United States), and from North Carolina he invited Clyde Hutchison III. To many people, this was yet further evidence of Mike's generosity.

Along with Elizabeth, Mike also invited Helen, their children Wendy, Ian, and Tom (and their partners), and his brother Robin, adding another eight people to his entourage and boosting his generous image. Before leaving they enjoyed notoriety of a different sort. Although Mike's wardrobe was famous for its casual charm, he had become increasingly accustomed to jackets and neckties as the occasion dictated. Now, however, nothing less than a formal, white tie tuxedo would suffice for the Nobel ceremony. Vancouver's social leaders eagerly gave Mike and Elizabeth fashion advice and featured them on the front page of the Style section of *The Van-*

couver Sun newspaper. Colleagues later joked that they feared the introduction of a new standard of laboratory dress. Local designers competed to provide Elizabeth with a wardrobe for the Stockholm event and other official visits, providing her with seven long dresses for white tie events and five short cocktail-length outfits for black tie events. Helen and Wendy did not escape the attention of local designers either. They opted for outfits featuring designs based on traditional Haida symbols.[8]

Mike was pleased with his personal success in winning the Nobel Prize, but he felt a responsibility to use his new status to promote and support high-quality biomedical research at UBC and across the country. He used the extensive media coverage of his win to publicize his worries about science funding in Canada as the country grappled with a national recession and capped MRC grants. He cautioned against an over-emphasis on "strategic" research, noting that most important scientific discoveries — like his own — were not planned. With a change in federal government on October 25 came promises to renew the national Networks of Centres of Excellence program, and Mike hoped that his prize would help convince politicians to provide more money for science funding in general.[9] Indeed, Mike's new status later provided opportunities to influence science policy at the highest level. He was only the fourth Canadian working in a Canadian laboratory to have won a Nobel Prize (along with Frederick Banting, Gerhard Herzberg, and John Polanyi), but other Nobel Prizes had gone to expatriates or foreign scholars working in Canada; perhaps Mike could increase the chances that other scientists in his country would win this important prize. At the provincial level, he noted with regret financial cuts to the British Columbia Health Research Foundation.[10]

Even before he left for the Nobel ceremonies Mike pulled a few political strings when he donated all his prize money to charitable causes. Half would support post-doctoral fellowships in schizophrenia research, a disease that had interested Mike for several years for very personal reasons. He had seen first-hand how frightening the disease could be and how poorly understood it was by the public and medical science. The other half of his prize money went to the Vancouver Foundation to fund public science education through Science World and the Society for Canadian Women in Science and Technology (SCWIST). A few colleagues were surprised by Mike's support for women — they could not remember him speaking up at department meetings to support women in faculty positions — but others accepted his rationale that he had always respected women as scientists and recognized that they encountered obstacles not faced by men. His first few doctoral students had been women, as were several of his most important collaborators. Behind the scenes, Mike had supported women as candidates for teaching and research positions in his department, although gender parity was still a long way off.[11] These gifts earned Mike the thanks and admiration of people with schizophrenia and their families, as well as women in science and anyone connected with Vancouver's Science World. Thank-you notes poured into the office from around the country, and Mike responded personally to every one. He even responded politely to critics of biomedical research who suggested alternative approaches to addressing schizophrenia.[12]

The politics of philanthropy intensified as he challenged governments to match the donations. Through contacts in the civil service he asked the province of British Columbia to match the $500,000 and then asked the federal government to match the combined million. The federal government offered a matching $500,000 but Mike insisted on more. In the end, Industry Canada (formerly Industry, Science, and Technology Canada) pledged $500,000 for science education and an additional $775,000 for research through the MRC, including $50,000 for an annual research prize (the Michael Smith Award for Excellence) and scholarships for graduate and post-doctoral research in schizophrenia. Mike was doubly pleased when the provincial government, having heard of the large federal contribution, contributed one million dollars to the Vancouver Foundation's Michael Smith Fund.[13]

Finally, after weeks of preparations and local congratulations, it was time for the trip to Sweden. Following tradition, the Nobel Prize was to be presented on December 10, the anniversary of Alfred

Nobel's death. Mike and Elizabeth left two weeks early to attend meetings in England, including a reunion of his Honours Chemistry class held at Oxford University. They also planned to visit people and places from his childhood, and to see a few shows in London before the week of festivities in Sweden. And what a week it was! Mike had a personal attaché and limousine in Stockholm, and he and his guests attended receptions, lectures, ceremonies, and cocktail parties every day. The Canadian ambassador to Sweden seized the rare opportunity to host a Canadian Nobel winner (and attend Nobel functions) and the Stockholm Symphony gave a special performance featuring Swedish citizen Barbara Hendricks, the American-born opera sensation and humanitarian. Along with the other Nobel Prize recipients for that year, Mike described his prize-winning work at a lecture on December 8. (Kary Mullis, who shared the 1993 Prize in Chemistry, gave a characteristically unconventional lecture that mixed personal with professional anecdotes — even his mother who attended was a little nervous about what he might say.)[14] On the day of the presentations all were whisked away to the Stockholm Concert Hall for the awards ceremony, Mike and Elizabeth in their limousine and his family and friends in special motor coaches. Local residents lined the streets to wave at the procession, making all the participants feel like celebrities.

Elizabeth and Mike's family sat in the front rows of the Concert Hall while others were further to the back and upstairs. The King of Sweden, Carl XVI Gustaf, and other members of the Royal Family entered ceremoniously to take their places on the right side of the stage shared with members of the Swedish Academy. Previous laureates who had chosen to attend also sat on the stage in an area known as "Penguin Mountain." The 1993 laureates entered escorted by young students and took their places at the front of the stage on the left for a ceremony that lasted more than an hour. After speeches (in Swedish) and musical interludes, the nine new laureates came forward to receive their gold medals and certificates from the King. Mike later joked that the ceremony was so well rehearsed that nothing could go wrong, unless he tripped or sat on his tails. He escaped any such embarrassment but he was nonetheless extremely excited. He could not stop grinning through the presentation of his award and his enthusiasm led to one minor but excusable *faux pas:* he initiated the handshake with the King. Friends in Vancouver who had gathered at Science World to watch the live telecast cheered, congratulating him once again during the interactive teleconference following the ceremony. To Mike, the whole event was awe-inspiring and a lot of fun. His only regret was that three key people in his life, his parents and Gordon Shrum, the man who brought him to Vancouver in 1956, were not alive to see him receive the Nobel Prize.[15]

A banquet at the Swedish City Hall followed, where the 1,300 guests were seated before the Royal Family and laureates entered. Mike, escorted by Princess Lillian, and Elizabeth, escorted by Prime Minister Carl Bildt, were seated with other distinguished guests at the head table with the King and Queen. Following toasts to the King and Alfred Nobel and songs from Barbara Hendricks came a sumptuous meal of reindeer, served on special Nobel chinaware used only for the occasion. Dinner concluded with a dessert served by members of a local men's choir who descended the staircase singing, illuminating their way in the darkened room with small lights under their platters. Mike gave an after-dinner speech as the representative of his Nobel Prize category in which he stressed the importance of protecting the earth's environment, even suggesting that the Nobel Committee might offer a new award to recognize excellence in this area. It was an audacious suggestion, but Mike was never afraid to give his opinions. The Canadians each took home gold foil-wrapped chocolate replicas of the Nobel medal as souvenirs, some of which were distributed upon arriving home in Vancouver, and Mike later purchased a set of white, gold-trimmed Nobel chinaware.

The dinner was just the beginning of the festivities. Mike stayed for the gala ball with Elizabeth and his family, thoroughly enjoying himself; at age sixty-one he still had plenty of energy for dancing. By midnight, Mike had had enough and he and Elizabeth left the others to continue the celebrations at a different venue hosted by the Medical Students at the Karolinska Institute. During the second party, Kary Mullis was reported to have climbed on a table where he mocked the Royal Family. Such behaviour did not win him favour and at the Nobel dinner in the Royal Palace the next evening, Mullis was seated as far from the Royal Family as possible. Mike felt an additional honour at the Palace dinner when he met the Nobel Peace Prize winners, Nelson Mandela and F.W. de Klerk, who had arrived from Oslo, Norway, where the Peace Prize is awarded. Mike later admitted how impressed he had been with these two men.

After several more days of regal treatment in Stockholm, Mike returned to Canada. En route to Vancouver, he and Elizabeth stopped in Ottawa as guests of the new Liberal government where they met Prime Minister Jean Chrétien, the President of the Medical Research Council, Henry Friesen, and personnel from other national science councils. They attended yet another gala ball, this time at the National Art Gallery. Prime Minister Chrétien was particularly interested in the Nobel medal and examined it for quite some time as photographers and news people covered the event. The media scrum that surrounded the Prime Minister immediately following the meeting was intimidating, but Mike was learning to meet with politicians and the press. Meetings with high-powered politicians and bureaucrats no doubt helped his cause to improve science research funding. The renewal application for his national network of protein engineers, PENCE, was under consideration, and early in the new year Mike reported to his network collaborators that he was confident of renewed funding.[16] Over the next few years, Mike would find additional opportunities to lobby in the name of scientific research.

If the fall of 1993 was hectic, so too was the entire following year. Mike could not say no to requests for his time, and there were many interruptions to his personal and professional plans. Mike gave seventy-four invited lectures in 1994 — an average of one every five days. Many were in Vancouver, but others were in Ontario, Quebec, Texas, Norway, Denmark, England, Italy, and Switzerland. He spoke at universities, hospitals, schools, and research institutes. He addressed scientists, business people, students, and the general public, lending his name to medical and social fundraisers. Some of the large events paid an honorarium, but many others did not and he sometimes returned any fees paid to him. Friends were impressed that he made as much effort to talk with students at Sir James Douglas Elementary School and elsewhere as he did with the scientific establishment. Mike took his visits with young people seriously, and he was amused when one little girl at a school science fair innocently asked whether he had ever done any research. He spoke to thousands of people about his science, the value of biotechnology, the value of science education, and the importance of basic health research.

Mike also travelled about the country to accept six major awards, including four honorary doctorates and induction into the Order of British Columbia. His philanthropy was recognized with four service awards, and a fifth in 1995. These honours were just the beginning. Mike continued to receive recognition from scholarly or medical societies in Europe and North America over the next few years. He acquired ten more honorary doctorates in 1995 (including one from the University of Manchester), became a Companion of the Order of Canada, and received another eight honorary doctorates in subsequent years. Mike was not simply a local celebrity, but a national hero.

Public appearances continued to take most of Mike's time during 1994 and much of 1995. He was constantly at lunches and dinners or sitting on a plane, following a hectic schedule that once had him giving eight talks in a single day. For a few months he showed people his gold Nobel medal, but switched to a bronze replica after someone dropped the original. He fraternized with the rich and famous, rode in a race car at the Vancouver Indy, and danced at the American Academy of Achievement Awards held in Las Vegas. Closer to home, he joined the Royal Vancouver Yacht Club thanks to the sponsorship of several friends who were already members, but he never purchased his own boat. (He did, however, avoid one public engagement in 1995 when he was excused from jury duty.) He was a Visiting Scientist to Taiwan in 1995, joining a Science Council of British Columbia biotechnology research and development mission. Mike lamented a sharp decrease in his skiing, boating, bicycle riding, and symphony concerts, although he managed to take holidays at his Bliss Landing home and the Langara Fishing Lodge, and a sailing adventure aboard the *Darwin Sound* to islands north of Norway. His passion for science remained as strong as ever, but he missed not getting into the lab and had to step down as scientific leader of PENCE. Even his personal life was on hold. He and Elizabeth, Mike's faithful companion on all his travels, waited until 1995 to move into a larger townhouse together and renew plans for marriage.[17]

During that first Nobel year, Mike received another distinction from his university. In July 1994, he became the Peter Wall Distinguished Professor of Biotechnology. The Peter Wall Institute of Advanced Studies was a UBC "think tank" launched in 1991 by President Strangway, who combined money from the Hampton Fund (supported by a new real estate development on campus) and local financier Peter Wall. This position permitted Mike's employment by UBC until June 2002, five years beyond normal retirement, and carried an additional salary. He was also permitted to fly business class when representing the university and bring Elizabeth. With his stipends as a Wall Professor and administrator combined with his MRC income, Mike earned one of the university's highest salaries at $175,000 per annum, plus a $50,000 expense account for university business. He was obviously an asset not only to the scientific community but to UBC as well. His appointment as University Professor, President Strangway had said, was a mark of the great esteem in which he was held by the university community, and was in recognition of Mike's "gracious and enlightened manner in which [he] handled the enormous public responsibilities which come with the winning of a Nobel Prize."[18] Mike, who still thought of himself as a socialist, may have felt a little embarrassed about his salary as he publicly commiserated with students who faced massive tuition increases and resented high administrative salaries.[19]

Public responsibilities eased a little after 1995, but never subsided completely. Invitations to speak at popular and academic scientific events — in Vancouver, across North America, in Europe, and in Asia — continued to arrive and Mike continued to accept unless he really did have a prior commitment. He explained that because his career had been supported by the Canadian public, he now had public responsibilities. Duty aside, he often seemed to enjoy all the attention. He spoke at medical fundraisers, academic symposia, and awards ceremonies for the prizes he had initiated, but he also gave his time to young people at local high schools. In all of his talks, Mike was positive and encouraging, supporting the values of his audience and avoiding controversy as was his nature. Biotechnology boosters heard of the vital social importance of their work and Mike's desire to eat a genetically modified tomato, medical researchers heard him advocate generous health research funding, science students — particularly young women — learned of good careers, and people with illness heard of hope through biomedical discoveries. Mike reminded old boys at Arnold School (where he opened the new Michael Smith Chemistry Laboratory) of the "good old days" but told British Columbia school trustees to resist the intrusions of private education. He praised participants in a Canada-Japan symposium on the stewardship of the north Pacific Ocean for their dedication, and he even told jokes to physicians about biochemistry classes at UBC. Through it all, he included in his talks a heartfelt appeal to take care of the planet and each other, identifying overpopulation as a major world problem. In addition to speaking in public, Mike responded to dozens of letters from autograph seekers (many from central Europe) and families coping with schizophrenia.[20]

Mike supported a few social causes, but he was very careful to choose those that fit his own values. He had in the past protested the removal of trees for new construction on campus and the industrial development of local wilderness areas. His trips to the Queen Charlotte Islands prompted him to lobby for the protection of South Moresby Island and the creation of Gwaii Haanas National Park Reserve. Over the years, he had joined a few peace walks as well, but now he lent his name to other significant issues. He joined the Canada Pugwash Group on nuclear disarmament and the Advisory Council of Science for Peace, although he had little time for active participation.[21] Mike became an Honorary Director of the British Columbia Schizophrenia Society, a Patron of Partners in Research, a Canadian educational group that promoted biomedical research and the humane use of laboratory animals, and he donated money from time to time to support other medical work or social services. He signed the Kyoto Climate Summit "Call for Action" declaration, a petition calling for reductions of greenhouse gas emissions, and another one protesting human rights abuses in Burma. He also wrote Prime Minister Jean Chrétien to support the many colleagues in biology who advocated habitat protection as part of new legislation to protect endangered species. Kofi Annan, Secretary of the United Nations, wrote to Mike as a Nobel laureate asking for his assistance if needed, and Mike was happy to consent.[22]

Mike even joined the Honorary Board of the David Suzuki Foundation in 1997 to support the environmental activism of his old friend David Suzuki. In a letter to Suzuki, Mike hinted — perhaps in jest — that his suggestion in 1993 to award a Nobel Prize for environmental protection may have helped with the 1995 Prize that recognized research into atmospheric ozone degradation. Mike thought that the David Suzuki Foundation sometimes sparked controversy and did not always have all the correct facts, but he was nonetheless pleased to lend his name, sign the occasional petition, and support a bid to elevate Suzuki himself to Companion of the Order of Canada. Representatives of the Suzuki Foundation were, Mike thought, a good counterbalance to those from politically conservative research groups such as the Fraser Institute.[23]

Still, Mike was careful whom he supported, applying two main criteria. The first was that advocates of the various causes had to present themselves clearly and politely. He refused to sign several petitions because he did not like how they were worded. Second, Mike would only aid a cause that had the support of internationally recognized experts. This was, he explained, to protect his own credibility and to ensure broader acceptance of the appeal.[24] He did not want to use his Nobel status to sway public opinion in areas where he did not have relevant expertise or the approval of those who did. The causes he supported had signatories who were leading scientists (including Nobel Prize winners), leading academics in other fields, or esteemed public citizens such as members of the Order of Canada.

Mike's regard for expertise made him reluctant to participate in some current political issues. For example, he refused to enter the conflict between British Columbia physicians and the B.C. Ministry of Health that controlled the province's public health care system because he was not an expert. Instead, he suggested to the physician who sought his support that "both the medical profession and government need to take a critical look at their mutual antipathy" to solve their differences. He eventually resigned from Physicians for Peace because he thought they lacked (or ignored) relevant expertise, and cautioned the Pugwash group not to exceed its mandate. Perhaps the only group Mike supported as a non-expert and which generated controversy was SCWIST, the organization to advance the careers of women in science and technology. Its affirmative-action approach to gender parity had many critics, but Mike continued to lend his name as an Honorary member "provided that it does not require me to undergo a sex change operation."[25]

He also drew the line when it came to criticism of biotechnology — the area where he had invested considerable time, energy, and money. As the 1990s wore on, Mike was surprised by the rising world-wide opposition to genetically modified organisms, especially in agriculture. As far as he was concerned, GMOs had considerable social value, caused no demonstrated harm to the consumer or the environment, and were the products of sound research in biotechnology. He thought that labeling, testing, and segregation of production lines might be inevitable to keep people satisfied, but were unnecessary and costly. Later, Mike joined a Royal Society of Canada expert panel to advise on biotechnology and lobbied the new Canadian Biotechnology Advisory Committee for a broader mandate to include plant and agricultural applications.[26]

Because of his role on scientific committees Mike was not about to defend GMOs publicly, and thought that industry support should come from the "true experts in plant and environmental science." He saw nothing scientifically wrong with GMOs, but regretted that agri-businesses had begun using the new technology to boost profits rather than increase food supplies. He thought that greed and arrogance had jeopardized the public acceptance of GMOs and was pleased to read about a large biotechnology company yielding to public pressure. Mike's acceptance of genetically modified organisms eventually led him to withdraw support for a David Suzuki Foundation declaration, believing that the Foundation had adopted a policy on GMOs with "no scientific analysis at all." Suzuki, a little upset that Mike would dismiss the Foundation's work in such a way, responded that the Foundation had taken no official position on GMOs and he personally advocated great caution with genetic manipulation, not total opposition.[27] As much as Mike believed in consultation and democratic decision-making, he sided with the experts whom he trusted when it came to evaluating biotechnology.

Despite the new demands on Mike's time brought on by the Nobel Prize, he still had routine work to do at UBC. The Biotechnology Laboratory needed some of his attention to set budgets, choose funding priorities, authorize expenditures, and resolve the earlier problems of grant "taxing," faculty teaching loads, and sabbatical leaves. However, two years elapsed between Mike's winning the Nobel Prize and a meeting of the advisory committee. When they finally met in April of 1995, it was clear that a new Director was needed. They prepared for an external review and made provision in the new budget for someone to replace Mike, who effectively stopped working as Director in the fall of 1996 but resigned officially the following year.[28] In the meantime, the Biotechnology Laboratory faculty who had been so carefully selected continued to do excellent research.

Members of Mike's own laboratory did not stop working while he was on the road, thanks in large part to a capable lab manager, Heather Merilees, and a highly independent group of research associates, post-doctoral fellows, and visiting professors. Two doctoral students finished their work in 1994, leaving one last student to finish, while others completed projects for PENCE and published their findings. The list of authors on the papers included Mike, but perhaps more as a formality than in recognition of his participation since many colleagues suspected that he had not been in close touch with his lab for several years. In 1994 Mike noted that he was growing older and suggested that "old farts such as myself" should retire as principal investigators and release space and money to younger scientists. But he did not want to end his career yet, and both UBC and the Medical Research Council were willing to provide a salary until 2002. He thought that he would take a sabbatical in 1996–97 and assess his skills, perhaps returning to UBC as a sort of visiting scientist.[29]

UBC also wanted to keep "its" Nobel laureate and included him on projects to enhance the institution's stature in the international academic community. Mike was a great supporter of interdisciplinary studies, and had earlier helped to organize Green College, an Oxbridge-style graduate residence at UBC established to mix students and visiting professors from various academic backgrounds. The Peter Wall Institute for Advanced Studies now began to plan interdisciplinary forums and research projects to bring diverse UBC academics together with visiting scholars, post-doctoral fellows, and graduate students. Mike was, of course, a valuable addition to the board. Their first sponsored project in "crisis point management" — to determine when activities or situations can be deemed at the verge of crisis — was criticized as overly utilitarian and not interdisciplinary enough: no historians, philosophers, artists, or representatives of different cultures. Still, the Institute pushed ahead with its most ambitious plan to wrest space from the old Faculty Club building that was being revitalized.[30]

As an active member of the university community Mike continued to express his views about policy, and not always in ways that pleased the administration. For example, he wrote to the Dean of Graduate Studies in 1997 to protest the introduction of differential fees for international graduate students. The provincial government had frozen student fees at 1995–96 levels and the university was missing a vital source of revenue. When the university President circulated a memo to confirm that higher tuition fees for international students were indeed under consideration, Mike slapped a note on it to say he should denounce the proposal as nonsense, "but I'm too annoyed right now." Similarly, he reacted strongly the following year when he heard that faculty were encouraged to use an airline company with which UBC had an "agreement." To Mike, this was an unacceptable encroachment on his personal freedom and a return to "the good old days of paternalism and autocracy." What would be next, he asked? Deals with automobile manufacturers and suggestions to buy certain cars? Penny-per-litre kickbacks from a petroleum company? Free vegetables for UBC Food Services in return for patronizing particular supermarkets? Ironically, two years earlier UBC had in fact signed a contract giving the Coca-Cola company a monopoly on UBC non-alcoholic beverage sales, and more agreements would come. These were all aspects of the same trends toward greater university commercialization that Mike had encountered earlier when searching for new research funds.[31]

Behind many of Mike's activities in the years following the Nobel Prize was an overarching concern to promote Canadian scientific research. His stature as a Nobel laureate provided special prestige to influence his academic field at many different levels, beginning with speaking engagements to promote public support for science and science education. Mike's financial investments were also having their desired effects. The Michael Smith Fund of the Vancouver Foundation was paying out to SCWIST and Science World by the spring of 1995, the latter providing inspirational science camps for school teachers among its projects. The Medical Research Council began awarding the $50,000 Michael Smith Award for Excellence in 1994, and by early 1996 Industry Canada was conferring "Michael Smith Awards for Science Promotion," although Mike had no role in their administration. Two of the first awards went to the nature magazine *Owl* and CBC Radio's science program Quirks and Quarks.[32] Mike's contributions to the MRC went further in 1996 when the funding council began providing its graduate studentships and, with financial partners, the Michael Smith/MRC/Ciba Geigy Professorship in Neuroscience and the Schizophrenia Society of Canada/MRC post-doctoral fellowships.[33] (Gobind Khorana wrote to Mike in 1998 about the possibility of an MRC Michael Smith research fellow coming to his lab). Mike was often asked to participate in reviewing candidates or presenting the awards, and he often did.

Another way Mike guided biomedical research was by serving on the boards of various scientific organizations. Mike had adjudicated grant proposals and scholarship candidates before the Nobel Prize, but he had more often shaped his field through editorial boards or as a scientific advisor. Membership on editorial boards lapsed following the Nobel Prize but he continued to review applications for Gairdiner Awards (in 1998 he refunded his 1986 award, plus interest, to support the Gairdiner Foundation's celebrations that year) and began to review grant applications for the Edith and John Low-Beer (EJLB) Foundation, a private body in Montreal that funded schizophrenia research. Mike was also in demand from private sector biotechnology companies. He joined the Boards of ICN Pharmaceuticals in Costa Mesa, California, and the Connaught Laboratories in Toronto, and the scientific advisory committees of Inex Pharmaceuticals (a UBC spinoff company) and Theratechnologies Inc. He also consulted with the MDS Health Group, a chain of clinical laboratories operated by a Vancouver physician.[34] Although Mike's role was scientific, he was not averse to participating in various organizations — charitable or commercial — to support biomedical research and the distribution of their products.

Part of what encouraged a self-described socialist like Mike to support charitable and commercial investment in science were the ongoing reductions in government funding for scientific research that had begun many years earlier. Mike was disappointed that the federal budget of February 1994 did not provide more money for the research councils, commenting that whatever strategic research was done in the Networks of Centres of Excellence depended on a foundation of basic, investigator-initiated science. At stake were Canada's established research programs and a future that depended on recruiting and educating the best scientific talent possible, particularly young Canadians. Industrial research was necessary and desirable, Mike thought, but universities were the proper home of basic, peer-reviewed research of the highest order. Industry had no incentive to fund basic research whose commercial value and medical application might take many years to realize.[35] If government would not fund university research, however, Mike would look elsewhere.

As the federal Liberals reviewed federal science policy in preparation for the 1995 budget, rumours of massive cuts circulated among scientists. Mike wrote the Finance Minister, Paul Martin, to lobby for the least possible reduction in science funding and later met with him while Montreal. Unfortunately for the science community, the budget of February 1995 presented an austerity program that slashed budgets for research in the name of deficit reduction. The Medical Research Council responded to its ten percent operating budget cut by reducing all its grants by five percent. The new science policy announced the following year brought further cuts to the MRC budget, leaving a funding void to be filled by other sources. Although the MRC escaped with lighter cuts than other councils — the Networks of Centres of Excellence were slated to be eliminated — retrenchment was drastic and caused considerable alarm in the biomedical research community.[36]

Scientists across Canada rallied to oppose the cuts, launching a successful public relations and lobbying campaign to save the Networks program which was made permanent in 1997, although new policy dictated that Networks had to become self-sufficient within fourteen years. Mike again contacted the Minister of Finance in early 1996 to discuss research funding, meeting with him briefly later that spring. Government officials were willing to listen to someone of Mike's stature, and in the fall of 1996 he gained a special opportunity to influence policy when Prime Minister Chrétien appointed him to the newly formed Advisory Council on Science and Technology under John Manley. The Advisory Council was charged with finding ways to create more partnerships between government and the private sector and to foster private sector leadership in technological innovation.[37] Mike's colleagues urged him to use his standing to lobby for new funds for the research councils, which he did, but in his usual diplomatic and congenial way. When Prime Minister Chrétien visited the Advisory Council in the fall of 1996, Mike presented the case for a long-term commitment to fund basic, curiosity-driven academic research. Private and charitable partnerships were fine, he argued, but universities had a special role to play in the unpredictable world of scientific discovery that preceded commercial exploitation. Furthermore, the federal government was best suited to take a leadership role. Mike repeatedly advised politicians that the research councils needed several hundred million dollars each to compete with science programs in the United States and Japan.[38]

Mike and other scientists across Canada (including the Prime Minister's brother, also a biomedical research scientist) eventually saw hope for their cause. In early 1997 Paul Martin, as Minister of Finance, announced a new program called the Canada Foundation for Innovation (CFI) which would provide $800 million over five years for research and development infrastructure costs at universities, colleges, hospitals, and other laboratories. Mike and other academics across Canada applauded the initiative, although some had concerns that the funds were only available when matched by grants from private or charitable sources, an arrangement that favoured such areas as engineering and biotechnology. But it was a much needed infusion, nonetheless, and Mike was already encouraging matching funds from other sources. Mike thanked Martin personally and soon found himself on the Board of Directors of the CFI to adjudicate funding proposals. Later that year, the Finance Minister promised to restore MRC funding to 1994–95 levels which injected almost another $130 million over three years. Mike again wrote a personal thank-you note, adding that a better annual allotment would be triple the 1994–95 level, about $700 million, to maintain a per capita investment comparable to that of the United States.[39]

During the budget cuts of the 1990s, the President of the Medical Research Council had embraced the inevitable and called for more non-government research funding partnerships and a broader mandate to include various health services. One new program began in 1994 when the MRC sponsored the Canadian Medical Discoveries Fund, Inc. Mike joined with other academics and business people in September 1995 to help the Fund allocate venture capital for new, high-tech, medically-oriented companies. The Canadian Medical Discoveries Fund sold shares — Mike never bought any — and invested the capital in companies that developed pharmaceuticals, diagnostic tools, or medical instruments. The Fund supported companies created by Canadian academics, some of whom were at UBC, in an attempt to retain talented researchers in Canada and to commercialize research initially funded by the MRC. During the three years Mike was with the Fund, share value rose and prospects were promising. In cooperation with the MRC, the Canadian Medical Discoveries Fund established a prize in 1997 to recognize the most outstanding discovery in an academic institution that led to a commercial application. Mike advised on this prize.[40]

In 1998, Medical Research Council leaders proposed a new organization. The Canadian Institutes of Health Research was to be a network of academic, government, and industry researchers with ties to health care providers supported by funds from government, charitable, and commercial sources. Many in the health research community opposed aspects of the new associations — several scholars at UBC were particularly adamant that the CIHR should have no commercialization mandate — but Mike supported any move that would see new research funds made available as quickly as possible, writing to Health Minister Alan Rock to support the concept of the CIHR. Mike believed that peer review (to maintain high standards of academic inquiry) with accountability and utility (for possible commercial exploitation) was an acceptable compromise to ensure investment from various sources for an excellent health care system, a good environment to retain Canada's best young scientists, and spinoff potential for biotechnology and other industries. Of course there were legal and ethical issues; Mike's advice was to fund research on those topics, too, but basic scientific research was the first priority. He warned colleagues to be careful to avoid partnerships with aggressive industrial partners that might want to buy political support from academics.[41] MRC funding was slowly rising, and when the federal government accepted the CIHR proposal it promised additional funding increases.

Although Mike's political style was polite and congenial — if a politician would not listen to him he simply found another one who would — there was no question that he had taken a particular stand in his quest to find funds to support world-class Canadian health research. The late 1990s saw public outrage over a perceived erosion of Canada's publicly funded health care system as governments cut budgets and considered schemes to permit greater participation by commercial enterprises. Many Canadians felt that Medicare was under attack in the name of deficit reduction and private profit. Some thought the same about Canadian universities although the public outcry was less vocal. To Mike, mixing public with commercial and charitable interests was necessary to save and expand Canada's universities and biomedical research capability. To critics, these compromises and adherence to a hierarchical model of health provision was far from desirable. Opponents of the so-called biomedical model criticized the emphasis on pharmaceuticals and surgery — "chemicals and cuts" — and the neglect of social and environmental determinants of health like poverty and pollution. Politics such as Mike's kept power and profit in the hands of the medical elite.[42]

While Mike was helping to shape national science policy he also had a very practical scientific project, but not at his university. When he made clear his intention to retire as a principal investigator at UBC in 1996, members of his lab moved on to find other work leaving only two others to finish their research. Mike closed his lab that was now in the Networks of Centres of Excellence building with one last party in June, attended by many of his recent students, post-doctoral fellows, and "all the usual suspects." He had a wonderful time at the day-long party, attending symposia, listening to humorous odes, and visiting the pub. The lab had been much more than a workplace to him, and its closure left Mike saddened that "the era of camaraderie has come to an end." He sent thanks to the MRC and returned his unused grant money. Unfortunately, the festivities were overshadowed by the tragic death of Mike's last doctoral student who had been struck by a speeding motorcycle soon after completing her thesis. Mike was devastated and established the Marianne Huyer Memorial Fund in her name to award a prize to the best doctoral thesis in the Department of Biochemistry and Molecular Biology.[43]

Although Mike's lab at UBC closed, he was not abandoning research. He had plans to visit the University of Washington's Genome Center in Seattle for a year to study genomics, the sequencing and analysis of an organism's entire DNA, and had an administrative leave so as to retain his full UBC and MRC salary. He was both excited to return to the bench and nervous that, at age sixty-four, he was too old to be useful.[44] Maynard Olson, Director of the Center, had known Mike since the 1970s and welcomed him as a visitor although Olson also wondered if his old friend could successfully return to the bench. Work was supposed to begin immediately in the autumn of 1996 but speaking engagements, awards presentations, meetings, and a much needed holiday intervened. When early in the new year he was finally able to dedicate three or four days each week to his work in Seattle, Mike joined forces with a graduate student who introduced him to the lab and the other scientists who quickly developed a close rapport with the famous Canadian. By the spring of 1997 Mike was immersed in lab work: M13 libraries, template preps, sequencing reactions, gels, and other tools used in high through-put DNA sequencing during the 1990s. Before long, he began to contribute useful insight into several scientific problems.

In fact, Mike was delighted to realize just how much he could contribute to the sequencing of the nucleotides in human chromosome 7 and other projects. Once again, he was having fun. Mike was determined to sequence difficult sections of DNA that had previously defied easy interpretation, maintaining that unsequenced gaps could hide aberrations responsible for genetic diseases. He manually wrote out elaborate diagrams of highly repetitive nucleotide sequences and then designed custom primers "by eye" that exploited a slight variation in the generally monotonous patterns.

Olson and others were impressed with Mike's success in using old-fashioned, non-computerized methods. Because his research on the genome of *Pseudomonas aeruginosa* (a pathogenic bacterium) was co-sponsored by a private foundation and a commercial business, Mike (and UBC) had to waive ownership claims and rights to intellectual property. Laboratory secrecy was not new to him, of course, but he was nonetheless annoyed. He asked his secretary at UBC to file correspondence about intellectual property rights pertaining to his research "with other items to do with this stupidity."[45]

Aside from his experimental work, Mike made notes on the organization of the lab and its administration to consider how a similar, medium-sized genomics centre could be established in Canada. Ever since his work at Sanger's lab in 1975–76 sequencing the phiX174 virus genome Mike had wanted to bring genomics "home." During the 1980s, as techniques for sequencing DNA improved (most notably the refinement of the chain-terminator method), the United States and several western European countries (later joined by Japan) began planning a bold initiative to sequence the human genome — estimated then at over 100,000 genes — in what after 1986 became known as the Human Genome Project.[46] Mike and his colleagues in the Canadian Institute of Advanced Research were keen to have Canada play a part in the project and by 1989 were lobbying for government funds. To proponents of genomics, the Human Genome Project offered new possibilities for disease detection, treatment, and perhaps cure. The gene responsible for cystic fibrosis, for example, was discovered in Canada by Lap-Chee Tsui in 1989. Not everyone in the Canadian medical research community was convinced of the health benefits of genomics — some argued that poverty was a greater determinant of health — but Mike and his influential colleagues in the CIAR nonetheless asked for $25 million or more from the Medical Research Council of Canada. They later proposed a broader initiative to combine money from all the funding councils, charitable foundations, and industry, minding Mike's dictum that "there should be 'something in it for them.'" In the meantime, Mike joined with what became the Canadian Genome and Technology program to encourage Canadian scientists to pursue genome-related work. MRC funding in genomics was low, as a result of limited finances and divided views, but in 1996 the Council agreed to provide $5 million over five years for allied projects and encouraged matching funds from the pharmaceutical industry.[47]

Mike was finally given the opportunity to launch a major Canadian genomics laboratory by a research agency he had not closely worked with previously. His science had always had implications for cancer research, but he had only occasionally considered the disease directly. He had advised the National Cancer Institute of Canada on various scientific matters intermittently since the early 1980s but in the 1990s, perhaps because of his brush with melanoma in 1992, he grew more interested in supporting the efforts of local cancer research funding agencies. By 1996, as government sources of research funding continued to wane, Mike was aiding the fundraising efforts of the Canadian Cancer Society and the British Columbia Cancer Foundation. These activities brought Mike closer to local philanthropists whose donations to the Foundation helped launch an ambitious $100 million fundraising campaign to boost the research capacity of the B.C. Cancer Agency.[48]

One of Mike's interruptions while trying to work at the University of Washington in the fall of 1996 was a request by the British Columbia Cancer Agency to help recruit a top Canadian scientist as the Agency's Vice President, Research. The leading prospect was Victor Ling, whom Mike knew from the 1960s when Ling was a doctoral student at UBC. Ling returned to Ontario after graduation but kept in touch with UBC scientists throughout his career, returning in 1981 as a visiting professor in Mike's lab to study DNA sequencing. Convincing Ling to leave his secure post at the Ontario Cancer Institute and the University of Toronto for a new project in Vancouver might not be easy, so Mike summoned all his recruitment skills and offered to do what he could to make the proposal more attractive. Ling was interested in the offer that included cross-appointments to UBC, Vancouver General Hospital, and the Vancouver Health Sciences Centre, but he had one special condition: Mike would have to join him at the B.C. Cancer Agency.

Mike hesitated but agreed on two conditions: he would need a real role, not simply an honorary one, and the blessing of UBC. With this promise Ling joined the Cancer Agency late in 1996 and began organizing a genomics laboratory for cancer research under Mike's supervision with a $25 million grant available from the Cancer Foundation. Ling viewed cancer as a genetic disease and believed that genomics would provide valuable information on diagnosis, treatment, and even prevention. Mike now wanted permission from UBC to move his office, only just re-established next to the new Networks of Centres of Excellence offices on campus. He cautiously approached Barry McBride, the newly appointed Vice-President Academic and Provost, over lunch at a fancy restaurant. McBride, declining the wine, was initially reluctant to let UBC's star ambassador join a rival organization but in the end he agreed, however, feeling that Mike could be trusted to continue representing UBC and to facilitate cooperation between the two bodies. Although Mike moved his main office to the B.C. Cancer Agency near Heather Street and Twelfth Avenue, several kilometres from UBC, he remained a University and Wall Professor at UBC.[49]

To make the genome laboratory a reality required additional fundraising to ensure that the B.C. Cancer Foundation reached its $100 million goal. The medical research community in Canada was still divided in its support for large-scale genome sequencing, but Mike hoped that opponents would not upset his plans and that the Medical Research Council might provide some support nonetheless. With little confidence that the federal government would provide the necessary financial support, despite the recent announcements of new funding proposals, Mike eagerly sought private and provincial funds. He helped launch the B.C. Cancer Foundation's Millennium Campaign in the spring of 1997, lending his famous, influential, and now bearded face to advertisements. In the name of good fun and charitable donations, Mike agreed to a short stay in a mock jail cell to raise "bail" from friends and colleagues, but he did turn down a request to sing in a Celebrity New Wave-aoke™ performance, wishing the organizers of the event success nonetheless. Money could buy laboratories, expensive equipment, and supplies, but more importantly it could provide competitive salaries to attract a first-rate staff. Once again, recruiting the best people he could find became Mike's central mission.[50]

By the autumn of 1997, Mike's work at the University of Washington was completed and he could now turn all of his attention to the new genomics laboratory in Vancouver. The B.C. Cancer Agency was praised in the local press as the institution that had "saved" science in the province by providing Nobel laureate Michael Smith with the facilities to continue his work.[51] Other British Columbia laboratories at Simon Fraser University and the University of Victoria were noted as participants in genomics, but the Cancer Agency's Genome Sequence Centre was to be Canada's first dedicated laboratory and the first to apply its findings directly to cancer research. Mike's new Sequence Centre would also participate in the international Human Genome Project.

Although Mike was dubious that the federal government would fund his new laboratory, he was determined to take advantage of the recent government initiatives. The Canada Foundation for Innovation, announced early in 1997, invited letters of intent for funding requests and prepared for the first round of reviews that autumn. The unfortunate heart attack and death of the first CFI President left a vacancy in the administration that Mike, as a CFI director, suggested filling with his old friend Bob Miller, "a superb science administrator." However, Miller, had moved to the University of Washington a couple of years earlier and was not about to leave. Mike suggested another candidate, David Strangway, who had recently stepped down from UBC's presidency. Mike had not always agreed with Strangway's policies (particularly his decision to close the UBC Faculty Club), but in general the two had worked together well. Strangway was appointed CFI President, and by the summer of 1998 he and the CFI directors were considering a proposal for funding from the British Columbia Cancer Research Centre (the research unit in the Cancer Agency that would host the Sequence Centre). Mike, as a signatory on the proposal, knew that he had a conflict of interest as a director of the CFI. He insisted that this was unacceptable to the CFI and bad for the reputations of UBC, the B.C. Cancer Research Centre, and himself. Mike even hesitated to offer advice on assessment criteria for genomic research applications, but did so anyway. As on other occasions, Mike probably excused himself from any voting and discussion of his own proposal, and waited for the results.[52]

Although the CFI would only fund infrastructure costs amounting to forty percent of a proposal, other British Columbia sources were now available. The B.C. Cancer Foundation had funds to match, as did the Science Council of British Columbia and the British Columbia Health Research Foundation. In April 1998 the provincial government was set to announce a plan to match CFI funds, possibly up to $100 million. UBC invited Andrew Petter, the provincial Minister responsible, to the university to announce the government's intention; Mike was asked to make a speech of thanks, and to help ensure that UBC appeared "very, very grateful that they [the provincial government] have come to the table as a funding partner." Winning a CFI grant would mean that Mike's Genome Sequence Centre could begin with all the funds it needed, but unfortunately the CFI grant proposal was rejected.[53]

The lack of additional funding was only a temporary a setback. More disturbing was Mike's health that had taken an unexpected turn for the worst. He had been fairly healthy all his life so he was shocked to learn in August of 1998 that he had a rare blood disorder called myelodysplastic syndrome. This meant that his platelet and cell counts (red and white) were low, leaving him in no pain but with a severely compromised immune system. Mike travelled to Stanford University for tests and a second opinion, learning the grim truth about his illness: there was no known cure, treatment with bone marrow transplant was not done on people his age, and the condition would progress to acute leukemia within a few years. He abruptly cancelled many of his trips, declined invitations out of the province, and resigned from several of his national committees. He apologized and explained the reason that he was unavailable, and perhaps for the first time he learned to say no, although he did make a few exceptions. He devoted his professional energy to two projects: building a proper home for the Biotechnology

Laboratory at UBC and organizing the Genome Sequence Centre. His concern for cancer research having now become very personal, he was convinced that the new research projects at the B.C. Cancer Research Centre would lead to better understanding and treatment of the disease.[54]

Hopes for success in realizing Mike's goals heightened over the following year. Mike's lobbying — for biomedical research in general and genomics in particular — persisted and he had powerful allies in the business community. During the national "Health Awareness Week" in the autumn of 1998, at the same time as the Canadian Institutes of Health Research was proposed, Mike publicly praised the B.C. Cancer Agency for taking on the responsibility for funding genomics when the Canadian government would not. Perhaps his criticism carried more political clout than he realized, for the 1999 federal budget for health research saw an increase in funding for basic research and research training. Mike was also pleased with a personal reference in a parliamentary report by Peter Adams, Chair of the Government Caucus on Post Secondary Education. He thanked Adams for the acknowledgement, recommended an annual medical research budget of $800 million, and suggested support for a newly established funding council, Genome Canada, which Adams promised to keep in mind.[55]

Mike and Victor Ling submitted a new funding proposal to the Canada Foundation for Innovation that joined the B.C. Cancer Research Centre and UBC's Biotechnology Laboratory as the Centre for Integrated Genomics. Both UBC and the Cancer Agency promised to make the proposed project a top priority, find matching funds, and make provisions to assign intellectual property rights. The Genome Sequence Centre was to be the link between the two institutions. By the summer of 1999 the CFI had awarded $9.35 million to the Centre for Integrated Genomics for the Biotechnology Laboratory, and Mike's team prepared another CFI proposal for the Cancer Agency that would prove to be successful. An additional small infusion came when Mike won the Royal Bank Award, which provided a $125,000 donation to a charity of the recipient's choice; Mike chose the B.C. Cancer Agency. In the meantime, Mike had recruited Steven Jones, an expert in the new research field of computerized bioinformatics, and then ex-patriate Canadian Marco Marra, an expert in DNA sequencing and genome fingerprinting, who joined the temporary sequencing laboratory in Burnaby, a Vancouver suburb.[56]

For a brief time in the fall of 1999 Mike shifted his attention to the provincial government. The B.C. Coalition for Health Research claimed that the provincial government had reduced financial support for the B.C. Health Research Fund by fifty percent over the past decade, with a corresponding decrease in provincially supported researchers. At two dollars per capita, British Columbia ranked seventh in Canada for provincial support, well behind Alberta's thirteen dollar per capita investment. Researchers needed contributions from the province to match federal funds and boost declining Medical Research Council grants. During an address at a local MRC dinner attended by the provincial Deputy Minister of Health, Mike suggested a provincial investment of twenty dollars per capita (about $80 million). He later wrote the Minister responsible to protest the "disastrous" decisions that would handicap biomedical research in the province. In reply, Mike was assured that the government was committed to health but would allocate a mere $1.5 million for health research that year, "an amount that reflects the reality of the current financial environment."[57]

Ultimately Mike and the other boosters of genomics research in British Columbia were successful. The federal government announced a huge $160 million expenditure over five years in the summer of 2000 for five Genome Canada centres, including one in British Columbia. Mike and colleagues established Genome British Columbia to administer research contracts with Genome Canada, industry, and charitable agencies in such areas as agriculture, aquaculture, forestry, the environment, and human health. They also promised to support research into the social and economic implications of genomics, and to set policy for intellectual property rights. But, Mike insisted, intellectual property agreements should not interfere with the mission of universities and charitable agencies to publish research results publicly and to further scientific knowledge.[58] A new provincial government finally contributed to genome research with over $10 million for the Biotechnology Laboratory and nearly $5 million for the new Genome Sequence Centre. With additional funds from UBC, the Biotechnology Laboratory received some $27 million for new facilities to be named the Michael Smith Building. The B.C. Cancer Agency also neared its $100 million fundraising goal, thanks to some $56 million in federal and provincial contributions, including a successful CFI grant.[59] By the summer of 2000, the future of genome research in Canada was assured, as much by lobbying and fundraising as by scientific merit.

Mike had reached the pinnacle of an ever-expanding career. But there was something that impressed people even more than his stature as a scientist and public personality. Even though he often met with heads of state, business and cultural leaders, professional athletes, and, of course, elite scientists, he could still share a beer with his friends, chat with students and staff, or meet with the general public. He remained as personable and approachable after his Nobel Prize as he had been before. Visitors to the Cancer Agency were surprised to find that he had a tiny, inconspicuous office. Some 143 colleagues, family, and friends from across the continent celebrated Mike's sixty-fifth birthday in 1997 with a symposium and dinner party in his honour. Old friends and former colleagues sent warm letters of greetings. Guests circulated a card to congratulate "the world's nicest Nobel laureate" while another added, "Happy Birthday, it's been a pleasure sequencing with you." Still another wrote to Mike's irreverent side: "Hot dogs between your toes and French fries up your nose! That's what birthdays are all about." Even UBC's new President, Martha Piper, who had worked with Mike on the Prime Minister's Advisory Council for Science and Technology, praised him as a great Canadian hero not only for his good science but also for his inspiration as a teacher, colleague, and humanist. Mike was well known at local restaurants, especially his favourite, Bishop's, where he often brought guests or presented staff with a salmon from his recent catch for their private consumption. He usually ordered Alaska smoked black cod and was known as "a man who liked his food and wine, and who also liked things to be rather simple." He never forgot that his impressive career began with four and a half years working in Khorana's lab.[60]

There was, of course, a much more personal and private side to Mike. Every year since 1993 he slipped away to Langara Lodge in the Queen Charlotte Islands with Elizabeth and occasionally a few close friends. He took two more sailing trips on the *Darwin Sound,* first around Cape Horn in 1995 and then to Alaska in 1999 where, despite his illness, he was a calming presence during a violently stormy night. He spent a few weeks each summer at his Bliss Landing home on the Sunshine Coast where he and Elizabeth planned to spend more time once he truly had retired. There he read, paddled his boats, and set prawn traps. Unlike his Whistler chalet enjoyed by students and colleagues, his Bliss Landing home was quiet and remote, an eight-hour drive north from Vancouver with ferry rides across Howe Sound and Jervis Inlet. Friends often dropped by to enjoy his hospitality, share a glass of wine, and watch the sunset from the deck. Mike still took pleasure in the symphony, theatre, art, and English television comedies such as *Black Adder, Keeping Up Appearances, To The Manor Born,* and *The Last of the Summer Wine.* He had his grumpy moments, but he generally remained playful, fun-loving, and non-confrontational; the sensitive boy from Martin Moss was never far away.

Because of Mike's professional success and active lifestyle, many friends after 1998 were unaware that he had a serious illness. Naturally, once they heard about his poor health they were very concerned, offering suggestions about diets or other therapies that might help. He thanked them for their kind words and noted that the condition was stable for now — he would be around for a few more years. Except for some reduction in travel, he continued to work as hard and with as much determination as ever. But while skiing in January of 2000 he felt an unfamiliar and frightening shortness of breath, and by April he had a nagging cough and difficulty keeping food down. His new medication did not seem to be as effective as hoped, and his condition worsened. Still, he managed a trip to Lindau, Germany, for a meeting with other Nobel laureates, and a trip to England to visit friends. He conducted teleconferences with the various boards on which he still sat and joined with colleagues in the Royal Society of Canada to create a national think-tank to advise government, part of a movement to make the Society more relevant to Canadian affairs. These activities were in addition to the long hours spent pushing himself and his colleagues through the triumphant summer of 2000 that saw genomics become a part of Canadian science policy.[61] Although he continued to work hard, he was not well and an infection while at Bliss Landing that August proved difficult to control. Fearing the worst, Elizabeth pleaded with Mike to see his physician about his medication, but he insisted on waiting until an appointment later in September.

Finding new energy, Mike continued with meetings for genomics research and kept his speaking engagements, including one for the Royal Columbian Hospital Foundation in nearby New Westminster. By the end of September, however, it was clear that his health was declining. Because he was too weak to operate a vehicle safely, Elizabeth began driving him to his appointments and he cancelled lunches with friends — something he rarely did — because he just could not eat. Judith Hall, a dear friend, UBC colleague, and physician met Mike for a beer one Thursday evening and realized immediately that he needed prompt medical attention. Mike delayed because he had a speech to deliver to UBC alumni that Sunday, October 1, but when Sunday arrived he was in no shape to proceed. One of the organizers of the alumni event telephoned to change the format, but Elizabeth, knowing that Mike was too ill, intervened. Minutes later the telephone rang again, this time with a good friend and university colleague of Mike's who inquired about his health and offered to deliver the alumni talk. Relieved of his obligation, Mike tried to relax but had a tense day and a terrible night.

Despite his protests, Elizabeth took Mike the next morning to see his physician, who promptly had him admitted to the Vancouver General Hospital. There he was given a private room. It was anticipated that he would undergo seven to eight weeks of chemotherapy after which, all going well, he would be home for Christmas. Elizabeth stayed close through that Monday and Tuesday to help with personal requests and to provide emotional support, permitting a few carefully screened friends, including Mike's son Tom, to visit. But Tuesday night, Mike's condition suddenly worsened, and Elizabeth returned Wednesday morning to find him very weak.

A few hours later, Mike's physician had a new diagnosis and there would be no chemotherapy. It took several minutes to accept the inevitable, but then Mike thanked his physician and the hospital staff for providing the best possible care, telling them what a good life he had had. After a private moment with Elizabeth to calm himself, he had her telephone a few colleagues at UBC and the Cancer Agency, asking them to visit; he had some final advice and words of encouragement for them. He then called a few close friends to say hello, but unbeknownst to them he was really saying goodbye. Helen came with their son Ian while Wendy, who was in England, reached her father by telephone later that evening. The following morning, October 5, Mike's administrative assistant responded to his incoming e-mail with a message that shocked the academic community worldwide: Michael Smith had died during the night.

Michael Smith left behind an enormous scientific legacy. As a scientist of international renown, he helped raise the standard of biomedical research at the University of British Columbia and in Canada to unprecedented levels. His scientific achievements while working in a small lab with a few accomplished associates won him the admiration of colleagues at home and world-wide, while his support for national science councils helped expand the quality of research across the country. As biomedical research became more costly and more competitive, Mike ensured the future of high-calibre science at UBC by obtaining funding and new talent. As a participant in the politics of science he helped to convince politicians, philanthropists, industry, and the public to invest millions of dollars in biomedical research at UBC and across Canada. Although the social and political implications of his efforts will likely be debated for years, he played an important role in initiating new scientific activity of the highest order. Indeed, some of the young scientists he recruited to the Biotechnology Laboratory and then the Genome Sequence Centre (now the Genome Sciences Centre) have become world leaders in their fields. In recognition of his great contributions to science, Mike will be remembered on plaques in the Copp Building, the University Centre (the former Faculty Club), and on the new Michael Smith Biotechnology Laboratories at UBC. The Michael Smith Award for Excellence, the Michael Smith Fund at the Vancouver Foundation, and several Michael Smith Fellowships for medical research all testify to Mike's scientific legacy. More recently, in 2001, the Michael Smith Foundation for Health Research was established in Vancouver to help build a strong research environment in British Columbia, funding health scientists at all levels to increase the province's competitiveness for federal health research funds.

Mike also left a more personal legacy. A month after his death, nearly one thousand people gathered at UBC to honour his memory. Guests came from across Canada, England, the United States, and elsewhere, including family, scientists, politicians, students, and university staff. Hundreds more sent regrets that they could not attend. A number of Mike's closest associates spoke on his considerable attributes as a scientist, colleague, mentor, and friend. This was a time to remember his admirable character and there was no shortage of memories. People remembered his brilliant, penetrating, and wide-ranging scientific insight, his energy, and his dedication to science. They also recalled his humour, his congeniality, his fairness, and his modesty. He had, they said, set high standards for them as academics and as people. One of Mike's first recruits to the Biotechnology Laboratory, Brett Finlay, perhaps summarized it best: Mike left big Birkenstock sandals to fill.

7

REPROGRAMMING GENES: A CLOSER LOOK

In the history of science, Michael Smith will be remembered for developing site-directed mutagenesis. It was a simple idea on the surface, but, like other significant advances in science, this technology depended on the skilled work of a research team and many years of careful thought and experimentation by an international community of scientists. For those readers wishing a more in-depth understanding of the science, this chapter provides a detailed summary of the precedents for and applications of site-directed mutagenesis.

When Mike began his nucleic acid research, several key discoveries in the area of molecular biology had already been made. As early as 1868, Friedrich Miescher isolated deoxyribonucleic acids (DNA) and ribonucleic acids (RNA). It was not until the 1940s that another German scientist working in the United States, Erwin Chargaff, characterized the chemical composition of DNA and RNA. Over the next ten years, two key experiments identified DNA as the genetic material.[1] In 1944, Canadian Oswald Avery and Americans Colin MacLeod and Maclyn McCarty were studying two strains of the bacterium *Streptococcus pneumoniae* at the Rockefeller University in New York. One strain was pathogenic with a thick polysaccharide coat surrounding the bacterial cells, and the other was non-pathogenic without the polysaccharide coat. These scientists extracted DNA from the pathogenic bacteria and added it to the non-pathogenic bacteria, with the result that the latter also became pathogenic with a thick polysaccharide coat. They concluded that DNA encoded something that specified production of the polysaccharide coat; we now know it was a gene for an enzyme required for the polysaccharide synthesis.

In 1952 Alfred Hershey and Martha Chase performed a different type of experiment. These scientists were studying the growth (replication) of bacterial viruses, which are not free-living organisms and must enter a host bacterial cell (*E. coli,* for example) to replicate. Hershey and Chase knew that the bacterial virus they were studying consisted of DNA surrounded by a protein coat, but wished to identify which component carried the genetic information. They grew the virus in conditions where the protein was marked with a radioactive label (S^{35}) or where the DNA was marked with a different radioactive label (P^{32}). When they infected the host cells they found that only the P^{32}-labelled DNA entered the host cell, producing progeny virus particles after replication, while the S^{35}-labelled protein component remained outside the cells. Hershey and Chase thus concluded that DNA was the genetic material.[2]

In the early 1950's a group of scientists working at Cambridge set out to determine the three dimensional structure of DNA. Two of them, James Watson and Francis Crick, studied Chargaff's chemical data and the X-ray diffraction patterns of DNA fibres obtained by Rosalind Franklin and Maurice Wilkins in London. Watson and Crick also knew the physical properties of DNA, such as that heating a solution of DNA caused a reduction in the viscosity of the solution and increased the absorbance of UV light. They proposed a three-dimensional structure for DNA, the now familiar double helix. This ladder-like structure has two long, alternating deoxyribose-phosphate backbones with nitrogenous bases extending from each deoxyribose residue. Each deoxyribose-phosphate-base unit, or nucleotide, forms the basic building block of a DNA strand. Watson and Crick envisioned four bases, A (adenine), T (thymine), G (guanine), and C (cytosine) extending out from each half ladder. The A and T bases interact and the G and C bases interact to form the steps or rungs of the ladder. The two strands of the DNA molecule are identical except that the sequence of bases on the strands are opposite, or "complementary" to each other, and the orientation of the strands is opposite. Hence, in the adjacent diagram, the bases on one strand are labelled upside-down. Watson and Crick also determined that the double stranded DNA molecule exists in a right-handed helical form.[3]

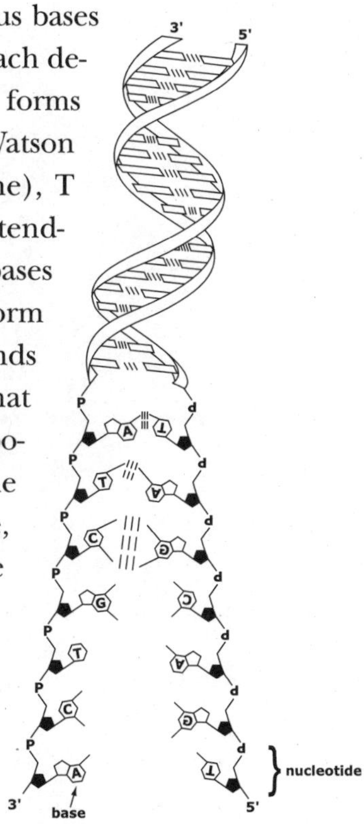

Watson and Crick's model for DNA captured the imagination of biochemists around the world, and it led to a flurry of additional studies to determine how DNA encodes proteins.

Other molecules closely related to DNA, ribonucleic acids or RNA, were also isolated and characterized. RNA contains an alternating ribose-phosphate backbone (not deoxyribose) and it is generally thought of as a single rather than double-stranded molecule, although double-stranded regions do occur that are important for function. The bases in RNA are A, U (uracil), G, and C, with U being similar to T in that it interacts with A. Soon, three main types of RNA were discovered. Ribosomal RNAs (rRNAs), the most common in cells, exist in RNA/protein complexes called ribosomes, the site of protein synthesis in cells. The next most abundant RNA in cells are the transfer RNAs (tRNAs) which transfer amino acids into protein chains being synthesized on the ribosomes. The third type of RNAs, messenger RNAs (mRNAs), constitute only about 2–5% of the total RNA in cells but are the links between the infor-

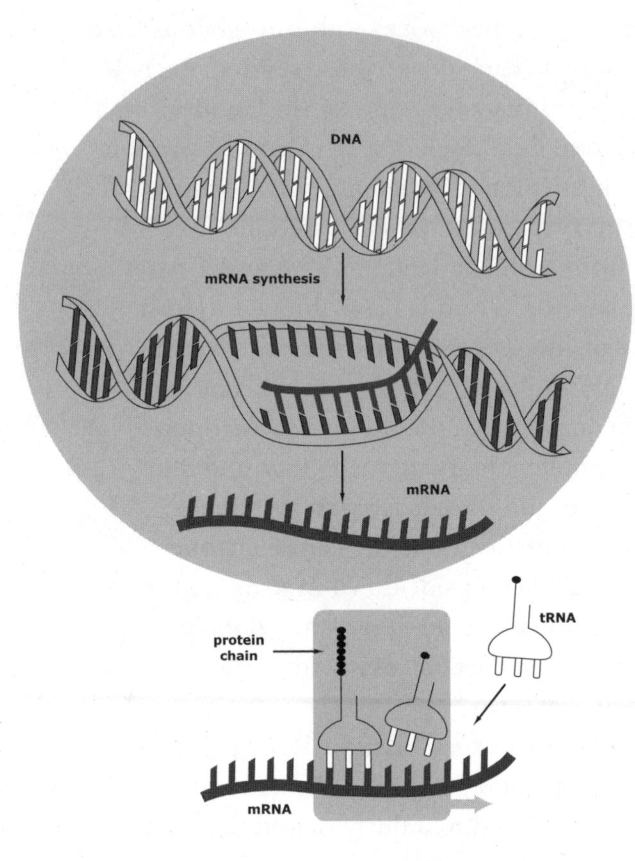

Information stored in DNA is copied into a mRNA and decoded into a specific protein.

mation stored in the genes (DNA) and the proteins. The sequence of the bases in DNA are copied into a complementary mRNA sequence which is decoded into a specific protein on the ribosomes. This knowledge gave rise to the central dogma of molecular biology that the sequence of the DNA specifies the sequence of the mRNA, which specifies the sequence of the amino acids in a protein.

How mRNAs encoded specific proteins was the next major challenge for biochemists. In the early 1960's several laboratories conducted a series of elegant experiments which determined that each amino acid was encoded by three bases on the mRNA. Scientist began to understand the genetic code, which is a triplet code. In addition, it was discovered that one triplet on the mRNA specifies where protein synthesis begins (the start codon) and three different triplets specify where protein synthesis ends (the stop codons). Marshall Nirenberg and Heinrich Matthaei added ribohomopolymers of either U, C, or A, to a bacterial cell extract and discovered that these polymers directed the synthesis of specific polypeptides, namely polyphenylalanine, polyproline, and polylysine, respectively. Hence the code for phenylalanine is UUU, for proline it is CCC, and for lysine it is AAA. Khorana's group, of which Mike was a member during his post-doctoral years, extended this work by synthesizing short stretches of DNA of defined repeating sequence and used an enzyme from bacteria to copy the DNA. Due to a "slippage phenomenon" the enzyme made long polymers of the same sequence which could be copied into an RNA polymer using an RNA polymerase. These ribopolymers also directed the synthesis of specific polypeptides.

Later, Nirenberg and Phil Leder used ribotrinucleotides (e.g. pUUU, pAAC, etc.) and an extract from bacteria that included the ribosomes (the organelles on which proteins are made) to show that these would form a complex with a particular tRNA that "decodes" the particular ribotrinucleotide. For example, $tRNA^{phe}$ can be coupled specifically to phenylalanine to give phe-$tRNA^{phe}$, the molecule that transfers phenylalanine into a protein chain. The phe-$tRNA^{phe}$ forms a complex with ribosomes in the presence of the trinucleotide pUUU. Therefore, the codon for phenylalanine is UUU. Similarly, asn-$tRNA^{asn}$ forms a complex with ribosomes in the presence of pAAC, hence the codon for the amino acid asparagine is AAC. Each of the 61 triplet codons for the 20 amino acids found in proteins was determined using a combination of these methods.[4] The Nobel Prize in Medicine was awarded in 1968 to Holley, Khorana, and Nirenberg for their interpretation of the genetic code and its function in protein synthesis.

It was an exciting time for a young scientist and Mike's interest in molecular biology was insatiable as he continued his own DNA studies while at the Fisheries Research Board laboratories and then at the University of British Columbia. However, by the late 1960s many of the fundamental properties and functions of nucleic acids were known when Mike began his work to understand better the interaction between short fragments of chemically synthesized DNA. When his doctoral student Caroline Astell needed a new project, Mike proposed a study to characterize the thermal stability of oligonucleotides bound to a complementary oligonucleotide or polymer. Mike wanted to know how well an oligonucleotide (or oligomer) of defined length and sequence could interact with a complementary oligomer or polymer when it formed a short region of double-stranded DNA due to the specific interactions of AT and GC base pairs. His motive was mainly curiosity, but he also wondered whether these oligomers could be used as probes to identify naturally occurring genes. Genes were of great interest now that they had been identified as the sections of DNA that actually determined the protein. No one knew at the time that this study would have great implications for subsequent developments. Many years later Mike acknowledged that these studies "started it all" (see below), meaning the road to site-directed mutagenesis and the Nobel Prize.

REPRINT FROM

LES PRIX NOBEL

1993

SYNTHETIC DNA AND BIOLOGY

By

Michael Smith

To Carol,

Michael

You started it all!

COPYRIGHT © LEX PRIX NOBEL 1993

Astell's experiments began by synthesizing deoxy-oligomers of a defined length and sequence using the phosphodiester method.

A, T, G, and C nucleotides, the raw materials, were coupled to form phosphodiester bonds using a type of chemical synthesis that was difficult and extraordinarily time consuming. For example, all but the T nucleotide had to be chemically modified to prevent unwanted side reactions from occurring during the synthesis of the bonds. This meant that synthesizing a trinucleotide such as dpApCpT included chemical protection of nitrogen residues on the A and C nucleotides (with Benzoyl (Bz) or Anisoyl (An) groups, respectively), and protection of the external phosphate with the cyanoethyl group (CE) and the external hydroxyl with the Acetyl group (Ac).[5] Thus to make dpApC, CE-dpA^{Bz} is coupled to dpC^{An}-OAc using a carbodiimide condensing agent, DCC (dicyclohexylcarbodiimide).[6] The coupling reaction was not particularly efficient, with rates of 50% at best.

Once the dinucleotide was formed the products were treated to retain the protecting groups on the bases but remove them from the ends of the dimer. This yields dpA^{Bz}pC^{An}. The dpT residue was added by protecting its external OH with the Ac group (dpT-OAc) and the external phosphate of the dpA^{Bz}C^{An} with the CE group (CE-dpA^{Bz}pC^{An}). Then these two intermediates were joined together, again using DCC. The efficiency of this joining reaction was also at best 50%. The desired products had to be purified on a column after each coupling step. With just a two step synthesis the overall yield was rapidly reduced (50% x 50% = 25%), necessitating large quantities of initial reactants — some 5 to 10 grams of each nucleotide — to synthesize a single oligomer.

The purification procedure required large columns of up to 2 litres in volume and as much as 16 litres of salt solution for washing. Because Mike's lab lacked a high capacity fraction collector, only small, 25 ml fractions could be collected, resulting in over 600 fractions per purification. These fractions were analyzed by measuring the amount of ultraviolet light absorbed by the solution: the higher the concentration of nucleotide, the more UV light was absorbed. Of course not every tube needed to be analyzed (maybe every 3rd tube) but it was all done manually as automated fraction readers were fairly new and expensive and Mike's lab could not afford one. The time consuming steps and low efficiency of the phosphodiester synthesis method meant that it could easily take two or three months, working seven days a week, to make an oligomer of approximately 9 residues in length such as dpTTCTTCTTC. However, the methods worked and within a year, enough different oligomers were synthesized to begin interaction studies between complementary oligomers.

In contrast to the chemical synthesis step, coupling the oligomers to an insoluble matrix material was relatively rapid and straightforward. Peter Gilham at Purdue University had devised a procedure to attach DNA oligomers to cellulose using a coupling reagent, this time a water soluble carbodiimide.[7] An aqueous solution of oligomer and carbodiimide was simply streaked onto a piece of cellulose paper and left in a moist atmosphere overnight. The efficiency of this reaction was about 25%, but a second application of the carbodiimide and another overnight incubation increased the efficiency by a factor of 2 (approximately 50%). Once the oligomer bonded to the cellulose paper, the paper was cut into small squares, stirred vigorously in a salt solution to create a pulp-like slurry, and poured into a small jacketed column custom-made by a local glass blowing company. By pumping water from a water bath through the jacket around the column it was possible to regulate the temperature of the column.

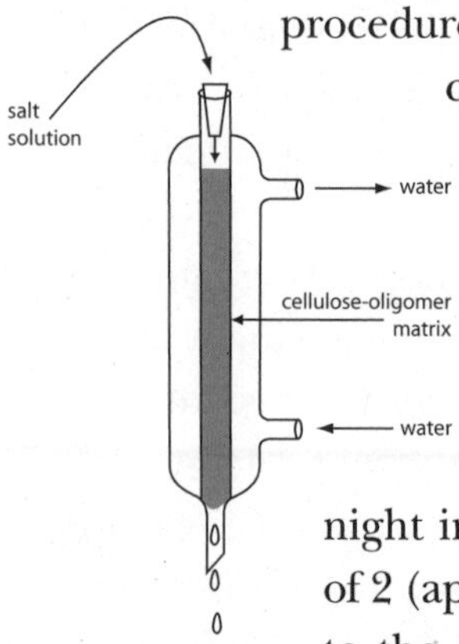

The moment of truth came when Astell added the oligomer $dp(A)_9$ to the first column containing cellulose and the complementary oligomer $dp(T)_9$. At about 4°C almost all of the $dp(A)_9$ stuck, forming a short double-stranded segment of DNA stabilized by AT base pairs. The temperature at which the complementary oligomers separated was determined by gradually increasing the temperature in the water jacket surrounding the column while continuing to pump salt solution slowly through the column. The $dp(A)_9$ oligomers washed off the column at a temperature of about 30°C, showing that the nine AT base pairs were no longer able to interact and form a stable double-stranded DNA molecule above this temperature and thus defined the melting or elution temperature (Tm) of the complex as 30°C under the conditions used.

A series of experiments using different complementary oligomers demonstrated that the Tm of a particular oligomer pair depended on its length, or the number of complementary base pairs that formed in the double stranded DNA fragment. Other factors were the number of stronger GC base pairs relative to the weaker AT pairs and whether the duplex involved two deoxyoligomers, which formed a stronger interaction than a deoxyoligomer/ribooligomer interaction. Other important studies showed that oligomer interaction also formed if there was a mismatched base pair in the synthetic DNA, although these interactions were less stable. For example, the figure below shows that dpA_8, which is completely complementary to the dpT_9 column, comes off the column at about 25°C while dpA_5TA_2, which would bind with a single mis-

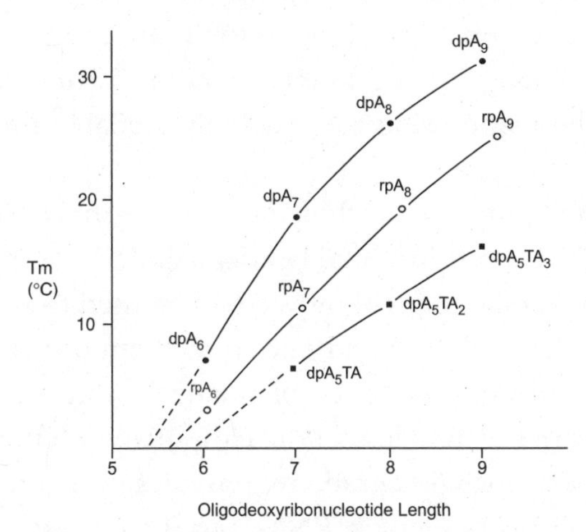

matched base pair, comes off the column at about 10°C. Thus the interaction between oligomers that have mismatches can be distinguished from those that do not by their lower stability. In general, a mismatch interaction resulted in a stability decrease of approximately 10°C.

So why were these experiments important? When this work began it was known that eukaryotes such as ourselves contained thousands of genes. (The current estimate is that there are 35,000 genes in the human genome.) Mike felt it should be possible to isolate a specific gene by attaching a fragment of synthetic DNA to a cellulose column to select a complementary sequence from the thousands of fragments of DNA derived from an organism's genome.[8] It should also work for selecting a specific mRNA, a task that was thought to be somewhat less daunting since the number of mRNAs expressed in cells was only an estimated 1,000 to 2,000. Assuming the bases in DNA are distributed randomly, one could estimate the approximate length of the oligomer required to be unique (occur once) within an organism's DNA genome. For example, the small bacterial virus phiX174 has a genome of about 5,000 bases. Since there are only four bases in DNA (A, T, C, and G), if they are distributed randomly then an oligomer that is 6 nucleotides long should occur once in that genome (4 raised to the power 6 = 4,096). Similar calculations for organisms with larger genomes indicated that oligomers 12 long should be unique for a genome the size of the common bacterium *E. coli,* 13 long for a simple eukaryote like yeast, and 16–17 long for the human genome.[9] These oligomer lengths could be made using the currently available phosphodiester chemistry.

Although Astell's project concluded very successfully, Mike questioned whether these results would be useful and where they would lead. Even if these oligomer columns could be used to isolate specific genes or mRNAs, how could one study them further? Procedures to determine the sequence of relatively long stretches of DNA or RNA were not available in the early 1970s. Methods using a number of specific RNA degradative enzymes were available for sequencing small RNAs such as tRNAs, but these were applicable only to relatively small RNAs that could be isolated in pure form in reasonably large quantities. Gordon Tener, Mike's long time colleague in the Biochemistry Department at UBC, had developed a method to isolate a specific tRNA (one of a mixture of 30–40 different tRNAs) from cells that could be sequenced using these specific degradative enzymes.[10] In contrast, mRNAs were present at much lower levels in cells, were much larger than tRNAs, and were much more heterogeneous with several thousand different molecules per cell.

For a brief period in 1971 while Mike was questioning where his DNA studies should go, he also considered switching his research to membrane proteins, a relatively new and exciting area of biochemistry. Membranes are the lipid envelopes that surround cells and many internal organelles, partitioning the cell into compartments such as the nucleus (where the chromosomes reside), the mitochondria (where cells generate energy to maintain life), and the endoplasmic reticulum (a complex series of interconnected membrane compartments that directs proteins and other molecules to specific sites within the cell, to the cell's outer surface, or to the external environment). Inserted into these membranes are specific proteins which facilitate the uptake or extrusion of molecules and relay signals from the environment into the cells. Mike wondered if he should begin to work in this area. To introduce himself to some of the techniques unique to membrane biochemistry, he took a six month sabbatical in Ed Reich's lab at the Rockefeller University in New York. However, at the end of the six months he returned to UBC and his studies of nucleic acids, including more studies on the interaction between complementary and mismatching oligomers.

By then Shirley Gillam, a very capable enzymologist, had joined Mike's group as a research associate and provided another important tool for the development of site-directed mutagenesis. Until 1970, the oligomers used in Mike's laboratory were made using phosphodiester chemistry methods. To simplify the procedure, many of the oligomers had repeating sequences. For example, the trinucleotide pAAG was polymerized to yield multiples such as a 6-mer dpAAGAAG, a 9-mer dpAAGAAGAAG, or a 12-mer dpAAGAAGAAGAAG. A short, complementary oligomer for a gene or mRNA would need to have a much more complex sequence. Mike suggested to Shirley that she try an enzymatic method. The enzyme polynucleotide phosphorylase (P.P.) purified from bacteria removed nucleotides, one at a time, from the RNA molecule, releasing a (ribo)nucleotide diphosphate (rppN).

$$\text{RNA} \sim\sim \text{pN}^{\text{x}}\text{pN}^{\text{y}}\text{pN}^{\text{z}} + \text{Pi} \xrightarrow{\text{P.P./Mg}^{++}} \text{RNA} \sim\sim \text{pN}^{\text{x}}\text{pN}^{\text{y}} + \text{rppN}^{\text{z}}$$

Even more intriguing was that incubation of the enzyme with RNA and a high level of diphosphate made the enzyme *extend* rather than degrade the RNA sequence. Furthermore, in the presence of manganese ion (Mn^{++}) rather than magnesium ion (Mg^{++}), the enzyme used (deoxy)nucleotide diphosphates (dppN) as a substrate, adding it to an existing piece of DNA.

$$\text{dppN}^{\text{z}} + \text{DNA} \sim\sim \text{pN}^{\text{x}}\text{pN}^{\text{y}} \xrightarrow{\text{P.P./Mn}^{++}} \text{DNA} \sim\sim \text{pN}^{\text{x}}\text{pN}^{\text{y}}\text{pN}^{\text{z}} + \text{Pi}$$

In no time Gillam had perfected the purification of P.P. and tried the experiment to extend chemically synthesized 3- and 4-mers enzymatically in a stepwise fashion. At each step the products were analyzed on a high pressure chromatography column and then purified on a larger preparative column. The efficiency of joining an additional nucleotide at each step varied from 30–50%, so in this regard it was no better than the phosphodiester methods. However, it was much faster, the nucleotides did not need to be protected chemically, and the scale of the reactions was much smaller so that purification was much easier.[11]

By the early- to mid-1970s molecular biologists were turning their attention to identifying genes and the proteins they encoded. Research on cloning DNA fragments and mRNAs was introducing useful methods to the nascent field of genetic engineering. Dan Nathans and Hamilton O. Smith at Johns Hopkins University had characterized a group of enzymes known as restriction endonucleases which cut DNA at unique sites determined by the nucleotide sequence. Paul Berg, Herb Boyer, and Stan Cohen at Stanford University had developed methods to clone DNA fragments by fusing them with small, self-replicating DNAs such as plasmids. This procedure was later modified to clone mRNA, actually a DNA copy of mRNA, called a cDNA (copy DNA).

At about the same time, other scientists heard of Mike's new methods to synthesize sequence-specific oligomers and contacted him to collaborate on projects that required these oligomers. One was Nobel laureate Fred Sanger at the Medical Research Council Labs in Cambridge, England. Sanger's lab was contributing to new technologies for molecular biology by developing an enzymatic method to sequence DNA based on replication of the DNA, in this case the phiX174 genome, at a specific region on the DNA template. As previously noted, phiX174 is a relatively small, bacterial virus (about 5,000 bases), unusual in that its genome is circular and single-stranded. During replication of the virus in bacteria, the viral genome is converted to a double stranded form typical of most DNA genomes. Sanger's sequencing method was known as the "plus-minus" method because the final reactions used two sets of test tubes with different conditions (labelled "plus" and "minus") where, making use of the different properties of two DNA polymerases, strand *degradation* stops at A, T, G, or C residues in the "plus" tubes while strand *extension* stops at A,T, G or C residues in the "minus" tubes, respectively. It was the first enzymatic procedure developed that allowed determination of the sequence of a long stretch of bases in DNA.[12]

Sanger had encountered regions of the phiX174 DNA for which he and his colleagues were unable to prepare a suitable short fragment needed to prime DNA synthesis. He thus contacted Mike to ask if he would chemically synthesize a short oligomer that they could use for this purpose. Mike was pleased to contribute but was also curious to learn more about sequencing DNA and so arranged a sabbatical leave in 1975 to work with Sanger on the phiX174 sequence project. Mike reasoned that it would be useful to learn how to sequence genes if his oligomer-cellulose columns could be used to purify them.

Mike devoted long hours in Sanger's lab to learn the plus-minus sequencing method. He spent five months preparing primers and another five carrying out the plus-minus sequencing reactions, ultimately contributing to the first successful enzymatic sequence determination of an organism's entire genome.[13] Soon after, Sanger's laboratory modified the complex plus-minus method to the much simpler "chain terminator" method that required only four reaction tubes.[14] This procedure was later automated and used very successfully in the sequencing of the human genome.[15]

Learning to sequence DNA provided Mike with additional skills and knowledge to manipulate the molecule and identify genes, but the chance meeting with Clyde Hutchison III sparked the idea for site-directed mutagenesis. Hutchison described how his lab inserted fragments of DNA from a mutant virus (which was not infectious due to the fragmented nature of the DNA) into a host cell *(E. coli)* along with an intact viral DNA from a second non-mutant virus. The host cell took up the foreign DNA as expected, and the mutation was transferred into the intact genome of some of the infectious progeny virus.[16] (This procedure was a sort of "marker rescue" experiment.) Selection for the specific recombinant was complicated, however, and the method was limited by the few known mutations that were available.

Mike realized that instead of using mutated DNA fragments as described by Hutchison, he could use synthetic oligomers produced in his lab for insertion into *E. coli.* He believed he could create any mutation at any site of the viral genome by specifying the mismatching base, thus genetically engineering heritable changes in genes. However, although earlier research in Mike's lab had shown that short, synthetic oligomers with a mismatched base could form a stable complementary duplex in a test tube, this was less likely to happen inside a bacterial cell. Mike reasoned that if these interactions would not form easily within bacterial cells, they would not generate mutations efficiently when introduced into a bacterial cell with infectious single-stranded phiX174 DNA. Hence he felt it would be necessary to bind the synthetic oligomer to the phiX174 DNA in a test tube and use a DNA polymerase to extend the oligomer into a complementary DNA. This would result in a more stable, double-stranded DNA molecule for introduction into the host *E. coli* cells. When the DNA replicated, theoretically 50% of the progeny virus should contain the mutation. This was a simple but potentially revolutionary hypothesis that might lead to a general method to reprogram genes.

Mike's lab made oligomers that would mutate the lysis gene in phiX174. This gene encodes a protein the virus uses to lyse (break open) the host *E. coli* cells towards the end of the viral replication process, releasing a burst of progeny virus particles. The lysis protein did not need to be functional in the laboratory since exposure to low levels of chloroform had the same effect. Patricia Jahnke, Mike's technician, chemically synthesized a 4-mer (dpGTAT) for Gillam to extend using purified polynucleotide polymerase. They made two twelve-long oligomers, one to create a mutation, a single base change into the viral DNA (dpGTATCC**T**ACAAA, am- 12-mer) and one to reverse this base change back to the original sequence (dpGTATCC**C**ACAAA, am+ 12-mer). (The bold letter corresponds with the base to be changed.) The mutagenic oligomer corresponded with the non-coding strand of the viral DNA represented by the top line in the illustration below. The lower line is the coding strand of this double-stranded DNA region. Reading the lower

```
dpGTATCCCACAAA
  CATAGGGTGTTTpd
```

line backwards (right to left, since the strands of DNA run in opposite directions) reveals a small snippet of the sequence of the lysis gene protein: TTG encodes the amino acid leucine, TGG encodes the amino acid tryptophan, and GAT encodes an aspartic acid residue.[17] The mutagenic oligomer (am- 12-mer) was expected to change the TGG tryptophan codon to TAG, a stop codon. As a result, synthesis of the lysis protein would stop prematurely resulting in a much shorter and non-functional protein.

After learning to work with *E. coli* in Hutchison's lab, Gillam returned to Vancouver and bound the mutagenic oligomer to phiX174 DNA, extended the oligomer using DNA polymerase, and joined the ends of the complementary molecule with another enzyme, DNA ligase. The duplex DNA was then transferred into host *E. coli* cells, the virus replicated, and progeny virus released through exposure to chloroform. Wild-type and (presumably) mutant progeny virus were then grown on a plate of special *E. coli*

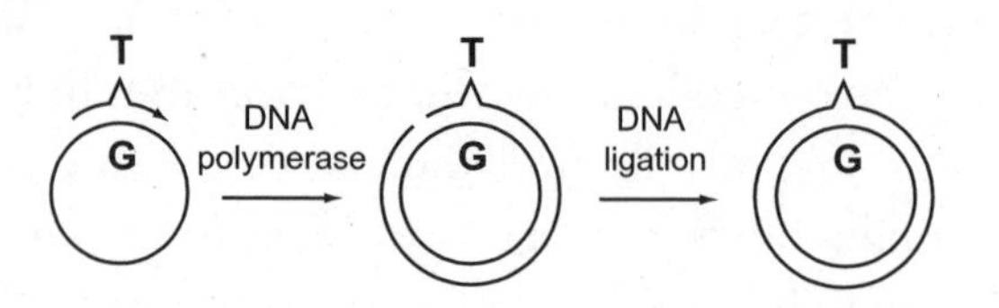

cells that carried a suppressor gene to allow protein synthesis to proceed past a TAG stop codon, generating a full length and functional lysis protein. Each virus that grew produced a small, clear, circular "plaque," a clearing in the lawn of bacterial cells. Each plaque grown from a single virus particle was sampled using a sterile toothpick and transferred onto two different plates of *E.coli.* The "C" plate had normal *E. coli* cells incapable of suppressing the TAG stop codon, while the "CQ2" plate had special *E. coli* cells that could suppress the TAG stop codon. If the virus grew on the "C" plate then it did not have a mutation in the lysis gene, but if it could not grow on the "C" plate, the virus should have the specific mutation.

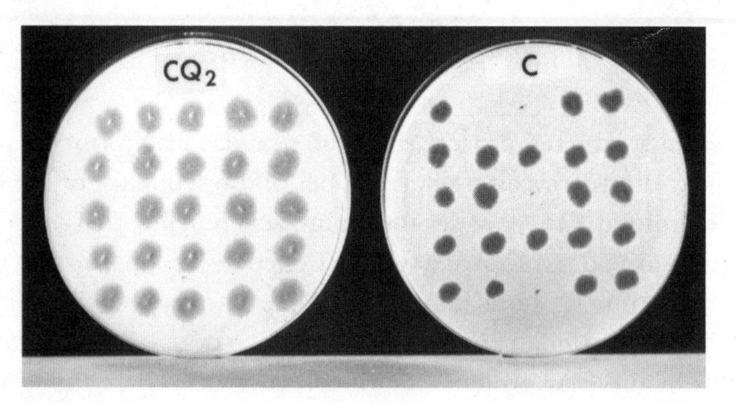

Identification of phiX174 virus with a mutated lysis gene by its inability to form a plaque on wild type Ecoli *cells (plate "C").*

The frequency of mutant virus recovered was extremely low in the first experiment. The team decided that the procedure to extend the oligomer into the complementary but mutated strand of phiX174 DNA was likely inefficient, leaving a mixture of fully and partially formed duplex molecules as illustrated below. The team decided to try eliminating the non-mutant phiX174 DNA from the

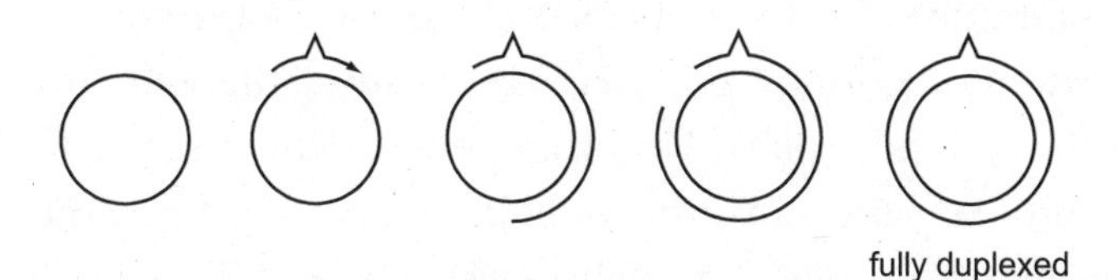

mixture by using another enzyme, a single-strand specific endonuclease that would degrade and inactivate any single-stranded regions of DNA leaving predominantly fully duplexed molecules. The yield of mutant virus recovered after repeating the entire experiment, including the endonuclease step, increased to a respectable 19 out of 225, or almost 8% mutant virus. Confirmation came after the am+ 12-mer reverted the mutant gene back to the wild-type form and "plus-minus" DNA sequencing showed the precise change expected in the DNA. The researchers had devised a method of site-directed mutagenesis using a small, synthetic oligomer as the mutagen.

While the method to change a base (nucleotide) within the lysis gene of phiX174 was very successful, much remained to make this a general technology that scientists worldwide could use to advance their studies. Different oligomers were synthesized to make other mutations, including the insertion or deletion of several bases at a time. Providing that a stretch of at least 6 nucleotides perfectly matched the template DNA at either end of the oligomer to ensure the specific interactions, the desired mutation was created. Also, using prior knowledge about the stability of oligomer interactions containing a mismatched base, Mike and Gillam experimented with different temperatures to improve the yield of recovering the desired mutation. For example, they used a mutagenic oligomer to bind to the first round of progeny virus DNA at a temperature some 10°C higher than that used to construct the mutation. Perhaps, they hypothesized, the mutagenic oligomer would bind preferentially to mutated DNA and during the replication process would selectively replicate the mutated DNA. This strategy worked very well. Two cycles of selection for the mutated DNA resulted in a significantly higher recovery of the desired mutation.[18]

Mike was also concerned about the practical limits of phiX174, a small spherical virus. Almost all of the phiX174 genome is needed for essential functions required for replication; there are no "extra" genes that can be replaced by other genes of interest (such as yeast genes, mouse genes, or even human genes). Gillam worked with Mark Zoller, a post-doctoral fellow who joined Mike's lab in 1981, to adapt site-directed mutagenesis for use with another bacterial virus known as M13. M13, also a single-stranded DNA virus, was proving useful to other researchers as a cloning vector that allowed easy insertion of any desired foreign DNA (known in lab slang as "YGI," your gene of interest). The M13 virus is filamentous, not spherical, and the length of the filament depends on the size of the viral genome which is naturally about 9,000 nucleotides long. Large insertions into the M13 genome simply extend the filament.

Zoller learned how to grow the M13 virus and soon adapted site-directed mutagenesis for use with it.[19] He also devised a method to simplify identification of mutated M13 virus. Zoller spotted DNA from individual M13 plaques onto filter paper and allowed a radio-labelled mutagenic oligonucleotide (the one used to make the mutation) to bind to the DNA. After washing the filter, the mutant viral DNA was identified because the mutagenic oligomer was a perfect match with this DNA and hence would form a more heat stable double-stranded DNA complex. When the filter was washed

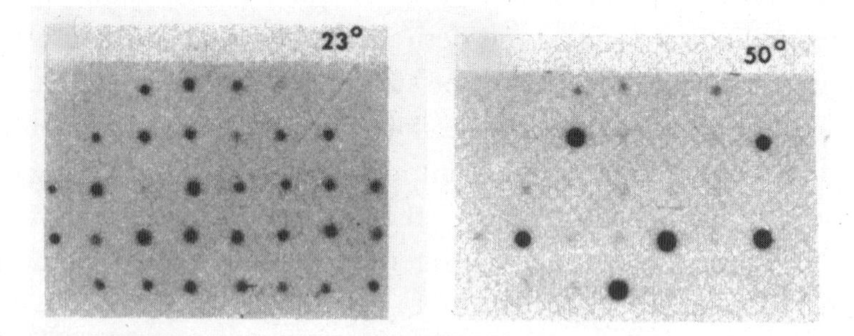

at higher temperatures, the radio-labelled oligomer remained bound to the filter corresponding with spots (viral DNA) that contained the desired mutation. A few years later, Zoller introduced another improvement to site-directed mutagenesis. He added a second oligomer along with the mutagenic oligomer used to generate a mutation. This second oligomer was a perfect match to the M13 genome and was positioned roughly half way around the viral DNA. This second oligomer increased the efficiency of complete synthesis of the complementary strand and thus increased the efficiency of recovering the mutant virus.[20]

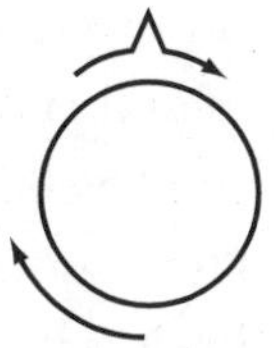

Later, other scientist also improved the efficiency of recovering a desired mutation. Two methods in particular worked well and were adopted as standard procedures in many labs. Thomas Kunkel used a DNA template that contained uracil (U) in place of thymine (T), a viral DNA that must be grown in a special strain of *E. coli.*[21] After binding the mutagenic oligomer and synthesizing the complementary strand, the DNA was transferred into a normal strain of *E. coli* which degrades the U-containing template DNA. In theory, only the complementary, mutated strand should replicate. Other researchers modified Zoller's "two-oligomer" technique but with a template DNA that contained a mutation to inactivate a selectable marker. The second oligomer mutated the selectable marker while the first oligomer created the mutation in the gene under study. As a result, mutated genes could be selected by simply selecting for growth, normally in the presence of an antibiotic such as ampicillin.

Site-directed mutagenesis was relatively simple, yet during the early 1980s it was recognized as extremely powerful. Scientists could create virtually any precise change anywhere in a segment of DNA to answer questions about the structure of the DNA and how it controlled expression of genes. By changing the bases of genes one altered the corresponding amino acid sequence and hence could ask questions about the structure, function, and mechanism of catalysis of proteins. Extending it further, enzymes could be modified to change — even improve — how they work. Mike thus had good claim as DNA's "mutagenius."

As the method proved sound, it was used to address important questions in biology. One of the first experiments in Mike's lab was to change a specific nucleotide in the gene for an enzyme that enables amino acids to form into proteins. The particular mutation converted an active site (the region on the enzyme where catalysis occurs) cysteine to a serine. The mutated protein was expressed at high level in bacterial cells and purified to allow biochemical characterization. This mutation lowered the activity of the enzyme by altering its affinity for ATP, one of the enzyme's substrates. Greg Winter, visiting from Cambridge University to collaborate with Mike and learn his methods, wrote in their paper, "We describe here a general method for systematically replacing amino acids in an enzyme. This allows analysis of their molecular roles in substrate binding or catalysis and could eventually lead to the engineering of new enzymatic activities."[22]

In a collaboration with Tony Pawson's group in the Microbiology Laboratory at UBC, site-directed mutagenesis was used to address the mechanism whereby the transforming protein ($p130^{gag\text{-}fps}$) from the Fujinami sarcoma virus is able to induce transformation of rat cells. At the time, transformation of cells was considered to be an important model system for study of the development of cancer. By changing a specific base in the gag-fps gene the scientists changed an amino acid, tyrosine, to phenylalanine in the GAG-FPS protein. The $p130^{gag\text{-}fps}$ protein is a tyrosine kinase and the mutation of tyrosine to phenylalanine reduced the activity of the enzyme fivefold and delayed significantly the transformed appearance of the cells.[23] These data supported the then current theory that tyrosine phosphorylation was important in regulating the oncogenic potential of this transforming protein. We now know that phosphorylation of proteins is an extremely important regulatory mechanism in cells and it plays a crucial role in the process of "signal transduction" which regulates, among other things, cell division and gene expression.

In other experiments Andrew Spence, a PhD student in Mike's laboratory, studied the role of sequences upstream of a herpes simplex virus gene. Steve McKnight from Seattle had observed that there were two GC-rich sequences upstream of the thymidine kinase gene, CCGCCC and GGGCGG. Since the sequences were inverted repeat sequences, they had the potential to regulate RNA synthesis by folding back and forming a hairpin structure. Did the hairpin structure regulate RNA synthesis? Spence made a series of mutations to alter these sequences either to prevent the hairpin structure from forming or to change the sequences without inhibiting formation of the hairpin structure. When the effect of the mutant constructs on RNA synthesis was tested, it was discovered that the sequence of the DNA and not the ability to form a hairpin structure played a role in regulating RNA synthesis. Instead of the earlier hypothesis, perhaps the GC-rich sequences might be binding sites for a protein that regulated RNA synthesis. Indeed, subsequent studies in the laboratory of Robert Tjian identified a protein that bound to the GC-rich sequences to promote RNA synthesis. This protein, called Sp1, was the first eukaryotic transcription factor to be characterized.[24]

Finally, Mike initiated a long-term collaboration in his own department that involved studies of metal-binding proteins that specifically bind metal ions to catalyze "electron transfer" reactions related to energy generation. These reactions help transport atmospheric oxygen to peripheral tissues or are involved in catalysis of specific metabolic processes. Mike's work in this area began when a postdoctoral fellow, Gary Pielak, joined his laboratory in 1983. Pielak had read Mike's work related to cloning and sequencing of the yeast *cytochrome c* gene and was interested in Mike's work on site-directed mutagenesis. Pielak thus proposed he apply site-directed mutagenesis to the study of yeast cytochrome *c*. Mike had a limited background in physical chemical studies of proteins so he suggested a collaboration with a faculty member in his department, Grant Mauk, who had a long-standing interest in physical studies of such proteins. Pielak soon began mutating the *cytochrome c* gene and preparing variants of the yeast cytochrome *c* protein in Mike's lab, and studying their spectroscopic and functional properties in Mauk's lab.[25]

Later in 1983, David Goodin arrived in Mike's lab as another postdoctoral fellow to work on expressing and mutating yeast cytochrome *c* peroxidase, a metal-binding enzyme that catalyzes the oxidation of cytochrome *c* by hydrogen peroxide. As variants of cytochrome *c* became available and their properties began to be understood, consideration turned to the effects that the amino acid substitutions might have on the structure of the protein. The definitive means of characterizing protein structures is by X-ray crystallography, so another faculty member in the Department, Gary Brayer, was brought into the growing collaboration to provide this experimental capability. Brayer's first graduate student, Gordon Louis, subsequently solved the 3-dimensional structure of wild-type yeast cytochrome *c* which had defied solution for many years. He also solved the structures of several cytochrome *c* variants produced by Pielak and Goodin. Other graduate students and postdoctoral fellows in all three groups also made significant contributions to this work. As the availability of these cytochrome *c* variants became known, researchers at other universities — including Robert J.P. Williams (Oxford University), Geoffrey Moore (initially Oxford University and then the University of East Anglia), and Brian Hoffman (Northwestern University) — established collaborative interactions with UBC.

Over the years, these studies helped to explain the functional importance of structural elements of cytochrome *c* that have been highly conserved during evolution, improved understanding of how cytochrome *c* peroxidase interacts with its substrate cytochrome *c*, and elicited the catalytic mechanism of cytochrome *c* peroxidase. With the organization of PENCE, this group of collaborators initiated a new research program with the long-term goal of increasing the limited catalytic activity of the oxygen-binding protein myoglobin through protein engineering, work that led to Mike's only patent. The collaboration of these three groups lasted nearly thirteen years and was brought to an end only by Mike's decision to close his laboratory at UBC in 1996.

Although site-directed mutagenesis was Mike's much-lauded breakthrough, his laboratory also played an important role in developing chemically synthesized DNA fragments (oligomers). The phosphodiester (chemical) methods and the combined chemical/enzymatic extension methods provided Mike's lab with enough oligomers to develop site-directed mutagenesis. By the early 1980s, Mike's capable research assistant, Tom Atkinson, was using a manual version of the newer phosphotriester method to synthesize oligomers. As site-directed mutagenesis proved important and widely useful, demand for oligomers from labs around the world grew at great rates. Most labs lacked the expertise to do their own chemical synthesis or enzymatic extension using polynucleotide phosphorylase, creating a huge commercial market for oligomers. Starting in the mid 1970's, Robert Letsinger at Northwestern University and Marvin Caruthers, first in Khorana's lab and then at the University of Colorado, played a major part in developing the phosphotriester method for synthesizing oligomers, a much more efficient method than the phosphodiester procedure. The phosphotriester methods evolved into a form in which the internucleotidic link is created as a phosphite triester, which is then oxidised to a phosphate triester. Caruthers and others also applied these methods to a solid-phase synthetic methodology. With demand for oligomers and a new and improved method to make them, an automated solid-phase instrument — a "gene-machine" — appeared on the market in the early 1980s.[26] An automated synthesizer could make an oligomer overnight whereas previously it often took up to several months. Mike and eleven other UBC faculty members pooled research funds to buy a machine that would make one oligomer at a time, and soon upgraded it to make three at a time.

These short fragments of chemically synthesized DNA were indispensable for site-directed mutagenesis and came to be useful in other applications such as the polymerase chain reaction (or PCR, a method used to amplify small regions of DNA useful for research and for clinical and forensic diagnostics). Synthetic oligomers were useful as linkers for cloning genes, as primers for automated DNA sequencing (hence playing an important role in the Human Genome Project), as antisense technology (a method to control the expression of genes in a cell), and many others.

By developing a method to reprogram genes, Mike had made his mark on the evolution of molecular biology. He was, however, the first to admit that he had help from the pioneers of the field who came before him, the community of scientists who contributed to his knowledge and understanding, and his lab colleagues who did much of the experimental work. Although it is always difficult to identify a single individual who was responsible for any given development, Mike, for good reason, was soon known as the "father of site-directed mutagenesis."

GLOSSARY

DEOXYRIBONUCLEIC ACID

A large molecule that determines the chemical sequence of proteins that constitute an organism, hence the genetic "blueprint" of the organism. It can be likened to half a ladder, with a backbone of alternating sugar (deoxyribose)-phosphate molecules as the ladder rail and a series of nitrogenous bases (nucleotides) attached to each sugar as half ladder rungs. The bases are either adenine (A), guanine (G), cytosine (C) or thymine (T). DNA exists normally as a double-stranded molecule (i.e. two half ladders come together to form the complete ladder-like structure). The two strands are held together by weak interactions between the bases. A always interacts with T and G always interacts with C. The weak bonds between bases (non-covalent interactions) permit easy separation for replication. The double stranded molecule is arranged in a right-handed helix. In cells, DNA is synthesized by separating the strand of the DNA molecule and enzymatically copying each strand, inserting the correct base according to specific base interactions, A with T, G with C.

ENZYMES

Normally proteins which carry out chemical reactions by reducing the activation energy of the reaction. In the absence of an enzyme, the reaction would occur very slowly if at all. The enzyme facilitates the chemical conversion by several orders of magnitude.

GENE

A region of DNA that encodes for a specific protein.

GENOME

The sum total of an organism's DNA. PhiX174, a small bacterial DNA virus, has just over five thousand nucleotides in its genome, while a bacterium such as *E. coli* has 4.5 million base pairs. The human genome contains over three billion base pairs.

NUCLEOSIDE

A nucleotide without the phosphate residue.

NUCLEOTIDE

The basic unit of DNA or RNA consisting of a sugar-phosphate molecule and its corresponding base (A, G, C, or T). A short string of nucleotides is called an oligonucleotide or oligomer. A much longer string is referred to as a polynucleotide or polymer.

OLIGOMER

An oligonucleotide or similar molecule; in laboratory slang, an "oligo."

OLIGONUCLEOTIDE

A segment of a nucleic acid; a contiguous series of nucleotides forming a section of half a DNA or RNA "ladder." Also referred to as an oligomer.

POLYMERASE

An enzyme responsible for synthesizing DNA or RNA. When the strands of a double-stranded DNA molecule separate for replication, each half ladder is replicated by a DNA polymerase enzyme recreating a double-stranded molecule. Both double-stranded daughter molecules are thus identical to each other.

PROTEIN

Molecules composed of amino acids that play important roles within cells. They may contribute to an organism's structure, regulate metabolism, provide immunity, permit motion, generate and transport nerve impulses, and serve other essential functions. Proteins are extremely important biological molecules. The sequence of proteins is encoded within the DNA. The information is copied into a messenger RNA which is decoded on the ribosomes to form proteins.

RIBONUCLEIC ACID

RNA is a molecule similar to DNA but includes an additional hydroxyl group in the sugar moiety and uracil (U) in place of thymine (T). It is also considered to be single-stranded although there are some double-stranded regions due to intramolecular base pairs. There are three main types of RNA in cells. Ribosomal RNAs are an essential component of the ribosomes, the cellular machinery responsible for synthesizing proteins. Transfer RNAs insert the correct amino acids into a growing protein chain. Messenger RNAs are copies of the DNA genes and encode the sequence of amino acids needed for synthesis of a specific protein.

SEQUENCING

The procedure to determine the order of constituent parts of a molecule. In the case of nucleic acids (DNA or RNA), sequencing determines the order of nucleotides.

VIRUS

A self-replicating molecule (DNA or occasionally, RNA, usually surrounded by a protein coat and in some cases a membrane) that invades a cell and makes use of the cell's biosynthetic mechanisms. A virus can mutate and adapt to changing environmental conditions.

NOTES

CHAPTER 1

1. John K. Walton, *Blackpool* (Edinburgh/Lancaster: Edinburgh University Press/Carnegie Publishing, 1998).

2. Most of the details of Smith's life in chapter one are from personal communication with family members and friends, and from Michael Smith's autobiographical accounts. See UBC Archives, Michael Smith Fonds, Box 1-4, biographical notes; Box 3-18, transcript of 1998 Arnold Lecture. See also www.nobel.se/chemistry/laureates/1993/smith-autobio.html; Lancashire Record Office, Blackpool Holy Cross baptism register, PR 3202/1/1. Mike's baptism certificate presently states that he was baptized Roman Catholic, which is inconsistent with other records and family accounts.

3. Christopher Martin, *A Short History of English Schools* (Hove: Wayland, 1979).

4. Ronald Webber, *Market Gardening: The History of Commercial Flower, Fruit and Vegetable Growing* (Newton Abbot: David & Charles, 1972).

5. E.J.T. Collins, "Rural and Agricultural Change" in *The Agrarian History of England and Wales 1850–1914,* ed. E.J.T. Collins (Cambridge: Cambridge University Press); Edith H. Whetham, *British Farming 1939–49* (London:

Thomas Nelson and Sons, 1952); John K. Walton, *Lancashire: A Social History 1558–1939* (Manchester: Manchester University Press, 1987).

6. "A Nobel Man," *The Blackpool Gazette,* 3 November 1995.

7. "St. Nick's School will be 100 this month," *Blackpool Gazette,* 3 January 1973; Kathleen Eyre, *Seven Golden Myles* (Clapham: Dalesman, 1961); Kathleen Eyre, *Fylde Folk: Moss or Sand* (Clapham: Dalesman, 1979); Smith Fonds, Box 2-29, 20 February 1995, Speech to Science World.

8. Ross McKibbin, *Classes and Cultures: England 1918–1951* (Oxford: Oxford University Press); Harold Perkin, *The Rise of Professional Society: England Since 1880* (London: Routledge, 1989).

9. Alan Calvert, *St. Nicholas Church: The First Hundred Years* (Blackpool: McVety and Taylor, 1973); Barry Shaw, "Research into the Origins of St. Nicholas School Marton Moss," (unpublished MS, in author's possession); John Lawson and Harold Silver, *A Social History of Education in England* (London: Methuen, 1973); Brian Simon, *The Politics of Educational Reform* (London: Lawrence & Wishart, 1974); Percy Patrick Hall, *A Hundred Years of Blackpool Education* (Blackpool: Blackpool Education Committee, 1970); Harold Monks, *Marton Moss and its Neighbourhood* (Blackpool: Cardwell, 1996).

10. A.J.P. Taylor, *English History 1914–1945* (Oxford: Oxford University Press, 1965).

11. P.H. Gosden, *Education in the Second World War* (London: Methuen, 1976).

12. Brian Simon, *Intelligence Testing and the Comprehensive School* (London: Lawrence & Wishart, 1953); Adrian Wooldridge, *Measuring the Mind* (Cambridge: Cambridge University Press, 1994). H.C. Dent, *Education in England and Wales,* 2nd ed. (London: Hodder and Stoughton, 1982).

13. Kenneth Shenton, *The History of Arnold School, Blackpool* (Preston: Carnegie Press/Arnold School, 1989). Not all private, proprietary schools earned state approval.

14. T.W. Bamford, *Rise of the Public Schools* (London: Thomas Nelson & Sons, 1967). Pennington's support for private education may not have been entirely altruistic. In 1920 he considered selling the school to the Blackpool Local Education Authority.

15. Raines Collection, Arnold School file, 4 September 1985, Smith to Rhodes; Bernard Crick, *George Orwell: A Life* (London: Martin Secker & Warburg, 1980); Robert Graves, *Goodbye To All That,* 2nd ed. (London: Cassell, 1958).

16. Smith Fonds, Box 14-27, excerpts from *The Arnoldian* 1942–1949; Raines Collection, Arnold School student reports 1943–1950.

17. Raines Collection, Arnold School reports 1943–1950; Oliver Sacks, *Uncle Tungsten* (New York: Alfred A. Knopf, 2001).

18. Michael Sanderson, *The Universities and British Industry 1850–1970* (London: Routledge and Kegan Paul, 1972).

19. Smith Fonds, Box 2-6, 25 March 1994, Transcript, Prime Minister's Awards for Teaching Excellence.

20. Raines Collection, Arnold School reports 1943–1950. Scholarship students from poor backgrounds were often steered into activities and social relationships that would move them from one sub-culture to another.

21. Henry Collis, Fred Hurll, and Rex Hazlewood, *B.-P.'s Scouts: An Official History of the Boy Scouts Association* (London: Collins, 1961).

22. John Springhall, *Youth, Empire and Society* (London: Croom Helm, 1977); Tim Jeal, *Baden Powell* (New Haven: Yale Nota Bene, 2001).

23. Raines Collection, 7 October 1950, Notice of Scholarship Award.

24. John Roche, "The Non-Medical Sciences, 1939–1970" in Brian Harrison, ed., *The Twentieth Century,* vol. 7 of *The History of the University of Oxford,* ed. T. Ashton (Oxford: Oxford University Press, 1994); Christopher N.L. Brooke, *A History of the University of Cambridge vol. IV, 1870–1990* (Cambridge: Cambridge University Press, 1993).

25. A.H. Halsey, "Oxford and the British Universities" in Brian Harrison, ed., *The Twentieth Century,* vol. 7 of *The History of the University of Oxford,* ed. T. Ashton (Oxford: Oxford University Press, 1994).

26. J.B. Morrell, "The Non-Medical Sciences, 1914–1939" in Brian Harrison, ed., *The Twentieth Century,* vol. 7 of *The History of the University of Oxford,* ed. T. Ashton (Oxford: Oxford University Press, 1994).

27. Smith Fonds, 5 February 1988, The University of Guelph Convocation Address.

28. H.B. Charlton, *Portrait of a University,* 2nd ed. (Manchester: Manchester University Press, 1952); V.H.H. Green, *The Universities* (Harmondsworth: Penguin, 1969); Brian Pullen with Michele Abenstern, *A History of the University of Manchester 1951–1973* (Manchester: Manchester University Press, 2000).

29. These are typical figures for redbrick universities.

30. James Mountford, *British Universities* (London: Oxford University Press, 1966).

31. Kingsley Amis, *Lucky Jim* (New York: The Viking Press, 1958). The original edition appeared in 1953.

32. University of Manchester, Department of Chemistry Archives (UMDCA), Register of Students volume 3, p. 860, DCH/1/2/2/3; Final Degree Register 1923–1954, DCH/1/2/4/3/1; Register 1933–1966, p. 221, 243, 263, DCH/1/2/3/3.

33. Christopher Driver, *The Exploding University* (London: Hodder and Stoughton, 1971).

34. UMDCA, Register of Students volume 3, p. 860, DCH/1/2/2/3; Quirks and Quarks, CBC Radio, 12 October 1996.

35. "Life of Research Rewarded," *UBC Reports* 39 (18), 28 October 1993, pp. 1, 5.

36. University of Manchester Archives (UMA), Reports of Council to Court 1953, part 2, pp. 55–58; UMA, Reports of Council to Court 1954, part 2, p. 59; William H. Brock, *The Norton History of Chemistry* (New York: W.W. Norton, 1993).

37. Smith Fonds, Box 14-3, newspaper clipping, *Globe and Mail,* 16 July 1994.

38. Alex Robinson, "Dr. Michael Smith and the path to the Nobel Prize," *Canadian Medical Association Journal* 150, no. 8 (April 1994): 1316.

39. Wilfrid Eggleston, *National Research in Canada: The NRC 1916–1966* (Toronto/Vancouver: Clarke, Irwin & Company, 1978), 284; UMA, Reports of Council to Court 1953, part 2, p. 56; http://www.nobel.se/chemistry/laureates/1986/polanyi-bio.html.

40. Gordon Shrum, with Peter Stursberg and Clive Cocking, eds., *Gordon Shrum: An Autobiography* (Vancouver: UBC Press, 1986).

41. Michael Smith, "Studies in the Stereochemistry of Diols and their Derivatives" (Unpublished PhD thesis, the University of Manchester, 1956).

42. Smith Fonds, Box 2-11, 1 July 1994, Canada Day Speech.

CHAPTER 2

1. Raines Collection, Order of Canada file, 17 January 1995, Smith to Owen; Patricia Roy, *Vancouver: An Illustrated History* (Toronto: James Lorimer, 1980), 146–7; Weather forecast, *The Vancouver Sun,* 22 September 1956, p. 8, and 24 September 1956, p. 10.

2. Gordon Shrum, "The B.C. Research Council," *UBC Alumni Chronicle* (1957): 22-23; "B.C. Research Council Holds Annual Meeting," *The Victoria Daily Colonist,* 31 July 1953, p. 2; B.C. Research Council, *Ninth Annual Report* (Vancouver: The Council, 1952–1953); "Overlapping In Research Work Claimed," *The Vancouver Sun,* 16 March 1949, p. 28; "B.C. Research Council Spurs Clash Between Minister, Coalition M.L.A.," *The Victoria Daily Colonist,* 19 March 1949, p. 27; "Council's Research Plan Wins Minister's Praise," *The Victoria Daily Colonist,* 8 February 1948, p. 3.

3. "New Research Building Opened," *The Vancouver Province,* 1 February 1952, p. 13.

4. Peter Waite, *Lord of Point Grey: Larry MacKenzie of UBC* (Vancouver: UBC Press, 1987); John B. Macdonald, *Chances and Choices: A Memoir* (Vancouver: UBC/Alumni Association of UBC, 2001).

5. UBC Board of Governors Minutes (hereafter "BoG Minutes"), 28 April 1952; http://www.nobel.se/medicine/laureates/1968/khorana-bio.html; Smith Fonds, Box 33-1, H. Gobind Khorana, "From Carbodiimide to Gene Synthesis," the XIIIth Feodor Lynen lecture, Miami Winter Symposium, 1981, 1-38; Shrum, *Gordon Shrum,* 71–72.

6. The huts were old army barracks moved onto campus following the war. Horace Freeland Judson, *The Eighth Day of Creation* (New York: Simon and Schuster, 1979), 488.

7. "BC Researchers Take 'Huge Stride' In Developing Cancer Fighting Drug," *Vancouver Province,* 19 August 1954, p. 1. Part of the embarrassment may have come from the inaccurate science reporting, a practice that has since improved. "Research Needs More Money," *The Vancouver Sun,* 26 July 1954, p. 4.

8. Michel Morange, *A History of Molecular Biology* trans. Matthew Cobb (Cambridge: Harvard University Press, 1998).

9. "$80,000 U.S. Grant Aids BC Scientist," *The Vancouver Sun,* 15 January 1957, p. 35; "Chemist wins $80,000 award for research into living cells," *The Vancouver Province,* 15 January 1957, p. 19; Faculty of Medicine Fonds, Box 1-16(2), 8 March 1956, Transcript, "Reply to the toast."

10. "Khorana team seeks science sponsor here," *The Vancouver Province,* 29 May 1959, p. 19.

11. R.A.J. McDonald, *Making Vancouver* (Vancouver: UBC Press, 1996); Jean Barman, *Growing Up British in British Columbia: Boys in Private School* (Vancouver: UBC Press, 1984), 163.

12. Eric Nicol, *Vancouver,* rev. ed. (Toronto: Doubleday, 1978), 186, 210–211; Michael Kluckner, *Vancouver the Way it Was* (Vancouver: Whitecap Books, 1984), 221, 231; Mike Tytherleigh, "Brewing," in *The Greater Vancouver Book,* ed. Chuck Davis (Surrey: The Linkman Press, 1997), 542–543; Robert A. Campbell, *Sit Down and Drink Your Beer* (Toronto: University of Toronto, 2000).

13. Jean Barman, *The West Beyond the West* (Toronto: University of Toronto Press, 1991), chap. 12; "Let's Spruce Up Vancouver Well Before the Centennial," *The Vancouver Sun,* 22 September 1956, p. 4.

14. Smith Fonds, Box 14-3, 23 June 1994, Smith to Rae. Mike claimed to have voted for the social democratic NDP party since arriving in Canada; Interviews and Personal Correspondence, 2002–3.

15. *B.C. Research Council Annual Reports,* 1952–1960.

16. The cost cited was the price in Canadian dollars in the late 1950s. "Scientists from BC win honors," *Vancouver Province,* 6 April 1959, p. 10; "Scientists acclaimed for life cell work," *Vancouver Province,* 7 April 1959, p. 7; *B.C. Research Council Annual Reports,* 1959, 1960.

17. Bog Minutes, 26 March 1956.

18. BoG Minutes, 3 June 1959, 6 July 1959.

19. BoG Minutes, 24 Sept 1945; Eric Damer, "The Study of Adult Education at the University of British Columbia, 1957–1985" (unpublished PhD thesis, the University of British Columbia, 2000); W.A.B. Bruneau, *A Matter of Identities: A History of the UBC Faculty Association 1920–1990* (Vancouver: UBC Faculty Association, 1990); Sandra Djwa, *Professing English: A Life of Roy Daniells* (Toronto: University of Toronto Press, 2002), 244.

20. http://www.nobel.se/medicine/laureates/1968/khorana-bio.html; *Calendar of the University of British Columbia* (hereafter *UBC Calendar*), 1959–1960. UBC's Department of Mechanical Engineering hired an Indian expatriate in 1961, but that was after MacKenzie's power had been broken.

21. "Federal aid sought for research centre," *Vancouver Province,* 11 June 1959, p. 6.

22. "Flower decked lawn scene of afternoon wedding," *Comox District Free Press,* 10 August 1960, p. 3.

23. Smith Fonds, Box 1-13, 17 January 1961, Tarr to Smith; 26 January 1961, Smith to Tarr; 10 February 1961, Idler to Smith; Box 33-1, 31 March 1961, Smith to Khorana.

24. Kenneth Johnstone, *The Aquatic Explorers: A History of the Fisheries Research Board of Canada* (Toronto: University of Toronto Press, 1977).

25. *Fisheries Research Board of Canada Annual Report 1961–1962* (Ottawa: the Board), 204.

26. Smith Fonds, Box 1-12, 10 February 1961, Idler to Smith.

27. Smith Fonds, Box 1-12, 9 March 1961, Tarr to Smith and 13 June 1961, Smith to Tarr.

28. Greg Myers, *Writing Biology: Texts in the Social Construction of Scientific Knowledge* (Madison: The University of Wisconsin Press, 1990).

29. Smith Fonds, Box 1-12, 7 September 1961, Smith to Tarr.

30. Smith Fonds, Box 4-8, 6 January 2000, Nakano to Smith.

31. Smith Fonds, Box 33-1, 4 September [1962], Khorana to Smith; 23 June [1963], Gobind to Mike; 25 March 1965 and [4 September 1962], Gobind to Mike.

32. Smith Fonds, Box 4-8, 6 January 2000, Nakano to Smith.

33. Smith Fonds, Box 33-1, 21 September 1964, Mike to Gobind.

34. Lee Stewart, *"It's Up to You": Women at UBC in the Early Years* (Vancouver: UBC Press, 1990); Lesley Andres Bellamy and Neil Guppy, "Opportunities and Obstacles for Women in Canadian Higher Education," in Jane Gaskell and Arlene McLaren, eds., *Women and Education,* 2nd. ed. (Calgary: Detselig, 1991), 163–192; *UBC Calendar,* 1959–1960; Faculty of Medicine Fonds, Box 13-16, President's Annual Report 1957–58, submission forms;

Box 15-2, 5 September 1963, Darrach to Bryce; Box 24-15, 18 December 1953, AMA Accreditation Survey (Form 9); Biological Discussion Club Fonds, Box 1, Minutes, 8 October 1925, *passim.*

35. Smith Fonds, Box 1-13, Research Projects [1965].

36. UBC Department of Biochemistry Collection, Smith Personnel File, 5 August 1964, Smith to Darrach, 7 August 1964, Darrach to McCreary, and 7 August 1964, Darrach to Tarr; Smith Fonds, Box 1-13, 7 August 1964, Darrach to Smith.

37. Department of Biochemistry Collection, Smith Personal file, 10 September 1964, Hoar to Okulitch.

38. Smith Fonds, Box 1-13, 10 February 1961, Idler to Smith.

39. Smith Fonds, Box 33-1, 21 September 1964, Smith to Khorana.

40. Smith Fonds, Box 33-1, 23 August 1963, Michael to Gobind.

41. Smith Fonds, Box 1-13, 7 August 1964, Darrach to Smith; 23 December 1964, Smith to Tarr. The UBC Financial Statements do not show Mike on the UBC payroll at that time.

42. Smith Fonds, Box 1-13, 30 July 1965, Smith to Martin.

43. Smith Fonds, Box 1-13, 3 August 1965, Smith to Tarr; Box 1-13, 30 July 1965, Smith to Martin. Mike transcribed Hayes' ultimatum. Smith Fonds, Box 1-13, 19 August 1965, Smith to Tarr. Mike transcribed Tarr's oral response.

44. *UBC Calendar,* 1959–1960; Smith Fonds, Box 1-13, 7 August 1964, Darrach to Smith; 11 March 1963; 3 May 1963, Tarr to Smith; Box 33-1, 23 August 1963, Michael to Gobind.

45. J.C. Stevenson, *Fisheries Research Board of Canada, 1898–1973* (Ottawa: Environment Canada, 1973); Larkin Fonds, Box 4-12, 23 September 1966, Larkin to Rogers.

46. Smith Fonds, Box 1-13, 20 July 1965, Hayes to Tarr; 2 August 1965, Memo from Hayes; 3 August 1965, Smith to Tarr; Smith Fonds, Box 22-25, 10 February 1966, Tarr to Associate Editor.

47. Smith Fonds, Box 22-21, 22 May 1964, Smith to Morgan.

48. Alvin Finkel and Margaret Conrad, *History of the Canadian Peoples: vol. 2,* 2nd ed. (Toronto: Copp Clark, 1998), 344, 345; *Medical Research in Canada: An Analysis of Immediate and Future Needs* (December, 1965); *A Legacy of Excellence 1960–2000* (Ottawa: Medical Research Council of Canada, 2000), 10.

49. Terrie Romano and Alison Li, *Celebrating the Medical Research Council of Canada: A Voyage in Time* (Ottawa: Medical Research Council of Canada, 2000); Faculty of Medicine Fonds, Box 15-2, 2 June 1959, Darrach to Weiser; 3 August 1961, Polglase to McCreary; *UBC Calendars,* 1950–1970.

50. Personnel File, 17 September 1965, Darrach and McCreary to Macdonald.

51. John B. Macdonald, *Higher Education in British Columbia and a Plan for the Future* (Vancouver: University of British Columbia, 1962); "Scientists: For Every One We Lack the Rank of Jobless Swells," *The Vancouver Sun,* 10 July 1965, p. 6; Smith Fonds, Box 33-1, 15 November 1965, Gobind to Mike.

52. Personnel File, 24 November 1965, Smith to Auer; Smith Fonds, Box 1-15, 26 January 1966, Wright to Smith; 2 February 1966, Smith to Wright; 23 March 1966, McCreary to Smith; 1 April 1966, Brown to Smith.

53. Personnel File, 24 March 1966, Darrach to McCreary.

54. Smith Fonds, Box 1-15, 23 March 1966, McCreary to Smith.

55. Smith Fonds, Box 1-12, 13 April 1966, Smith to Tarr; Raines Collection, Job Offers file, 13 April 1966, Smith to Schiff.

56. Smith Fonds, Box 1-15, 18 April 1966, Darrach to Brown; Personnel File, 24 May 1966, Darrach to Brown and 23 June 1966, Darrach to McCreary.

CHAPTER 3

1. Smith Fonds, Box 3-11, 27 October 1998, transcript, speech to the Vancouver Board of Trade.

2. Faculty of Medicine Fonds, Box 13-16, 1 October 1952, Annual Report to the President on the Faculty of Medicine; Harry Logan, *Tuum Est* (Vancouver: The University of British Columbia, 1958).

3. *Canadian Medical Research Survey and Outlook-MRC Report No. 2* (Ottawa: The Council, 1968); Personnel File, 20 October 1969, Dekker to Darrach.

4. Smith Fonds, Box 33-1, 21 January [1972], Gobind to Mike; 25 January 1972, Michael to Gobind; Box 1-19, 11 March 1970, Smith to Reich.

5. A modification of the G-methylation method was developed about 10 years later for use in the successful chemical method to sequence DNA. A. Maxam and W. Gilbert, "A new method for sequencing DNA," *Proceedings of the National Academy of Sciences* 74 (1977): 560–564.

6. Smith Fonds, Box 1-15, Report to MRC on research carried out 1968–73.

7. Department of Biochemistry, Smith Scrapbook File, 25 September 2000, Smith to Mackie.

8. The term "allosteric enzyme" was introduced to describe enzymes whose activity can be regulated. An enzyme is (normally) a protein that binds a substrate which is chemically altered due to this interaction. For example, phosphofructokinase converts fructose 6-phosphate to fructose 1,6-bisphosphate. This is a key step in glycolysis, a process whereby cells use glucose to generate energy. The substrate binds to a region on the enzyme called the catalytic site. Usually there is some 3-dimensional complementarity between the substrate and catalytic site, facilitating the interaction and contributing to the enzyme's specificity. In addition other ligands (small

molecules) may bind to other "allo" sites on the enzyme also due to some stereospecific complementarity between the ligand and these "regulatory" sites. The binding of the ligand may increase the rate of catalysis (positive regulation) or decrease the rate of catalysis (negative regulation). In this way, the activity of enzymes can be regulated rapidly, within seconds or less. In the example of phosphofructokinase, ADP increases the activity of the enzyme increasing glycolysis, while binding of phosphoenolpyruvate decreases its activity.

9. Personnel File, 20 October 1969, Dekker to Darrach; Christopher L. Hives, "The History of the U.B.C. Faculty Club: 1947–1970" (unpublished TMS, UBC Special Collections, 1984).

10. McCreary Papers, [January 1967], E.W.R Steacie Memorial Fund — Dinner for Dr. G.H. Dixon; Community Relations Fonds, Box 7-21, 27 April 1964, 12 April 1966, 3 Feb 1967 (2 items), News Release; Smith Fonds, Box 1-15, Report to MRC on research carried out 1968–1973.

11. Smith Fonds, Box 1-1, CV; Box 22-35, 2 April 1969, Gordon to Smith; Box 23-466 January 1969, Gordon to Smith; Box 24-13, 7 April 1977, Smith to Proudfoot.

12. The nucleus contains the chromosomes, the mitochondria contains an energy transducing system that generates energy to maintain life, and the endoplasmic reticulum is a complex series of interconnected membrane compartments that transfer proteins and other molecules to specific sites within the cell.

13. Smith Fonds, 8 May 1970, Smith to Darrach; 20 July 1970, Smith to Taylor; 14 August 1970, Taylor to Smith.

14. Personnel File, 11 June 1973, Polglase to Bates.

15. Smith Fonds, Box 1-15, 1 November 1966, Mason to McCreary; 15 April 1969, Stinson to Smith; 17 April 1969, Taylor to Smith; Progress Report on Research Studies [2 October 1968]; Personnel File, 11 June 1973, Polglase to Bates.

16. Faculty of Medicine Fonds, Box 20-10, 6 February 1984, Smith to Chalmers; Box 23-7, 16 May 1973, Flanagan to Bates; Box 23-7, Department of Biochemistry Review, [1973]; Smith Fonds, Box 1-13 Teaching Evaluation, n.d.

17. E. Gordon Young, *The Development of Biochemistry in Canada* (Toronto: University of Toronto Press, 1976), 36–37; Department of Chemistry Fonds, Box 3-10, Report on Biochemistry [1947]; Box 4-5, 5 July 1950, Hooley to Hedrick.

18. Faculty of Agricultural Sciences Fonds, Box 54-41, 26 September 1968, Ormrod to faculty curriculum committee; Faculty of Medicine Fonds, Box 24-23, AMA Report of the Survey of the Faculty of Medicine, 1970; Box 23-7,

Department of Biochemistry Review, [1973]; Box 25-1, Report of the Survey of the Faculty of Medicine, 1974; Box 26-1, CMA Report of Survey of the Faculty of Medicine, 1980;

19. Faculty of Medicine Fonds, Box 15-2, 8 October 1963, McCreary to White; Box 23-7, Department of Biochemistry Review [1973]; Box 25-1, Report of the Survey of the University of British Columbia Faculty of Medicine, 15-28 March 1974.

20. McCreary Papers, Box 9-3, 10 February 1967, Banks to McCreary; Box 10-2, 24 July 1967, transcript.

21. Faculty of Medicine, Box 13-16, various reports; Box 23-7, Department of Biochemistry Review [1973]; Box 15-2, various letters; Senate Subject Files, Box 28-35, 20 February 1958, Darrach to MacKenzie.

22. Senate Subject Files, Box 28-35, 20 February 1958, Darrach to MacKenzie; Faculty of Medicine Fonds, Box 15-2, 3 August 1961, Polglase to McCreary; 22 October 1963, Polglase to McCreary; Box 23-7, 24 August 1973, "An Analysis of the Review of the Department"; 1 June 1973, Bates to Volkoff.

23. Medical Research Council of Canada, *Survey of Research Personnel in the Medical Sciences in Canada 1965–66* (Ottawa: The Council, 1966).

24. Douglas Owram, *Born At The Right Time: A History of the Baby-Boom Generation* (Toronto: University of Toronto Press, 1996); Mark Thompson, *The Development of Collective Bargaining in Canadian Universities* (Vancouver: Reprint Series of the Institute of Industrial Relations, 1975); McCreary Papers, Box 9-3, 30 January 1967, petition.

25. Personnel File, 20 October 1969, Dekker to Darrach.

26. UBC Faculty Association Fonds, Box 4-7, 21 November 1969; November 1969, Report to the Personnel Services Committee; Box 5-7, [December 1970], Smith to Pearse.

27. Personnel File, 14 October 1969, Darrach to McCreary and 20 October 1969, Dekker to Darrach.

28. Personnel File, 22 October 1969, Darrach to Dekker and 31 December 1970, Darrach to McCreary.

29. Faculty Association Fonds, Box 4-7, 8 April 1971, Report of the Personnel Services Committee 1970–71; Box 4-8, 6 April 1972, Report of the Personnel Services Committee 1971–72.

30. Raines Collection, Job Offers file, 9 June 1973, Bocking to Smith.

31. Faculty of Medicine Fonds, Box 26-6, "Report to the Faculty of Medicine from the Committee on Faculty Organization"; Box 26-8, Resolutions of Senate.

32. Faculty of Medicine Fonds, Box 24-23, AMA Report of the Survey of The University of British Columbia Faculty of Medicine, March 1970; Box 26-6, 18 May 1971, Minutes; Box 26-6, 9 December 1971, Minutes; Box 23-7,

29 September 1972, Bates to Darrach; 16 May 1973, Polglase to Bates; 4 July 1973, Polglase to Bates. A rumour circulated at the time that Darrach also tendered a resignation that was immediately accepted.

33. Faculty of Medicine Fonds, Box 23-7, Department of Biochemistry Review [spring 1973].

34. Faculty of Medicine Fonds, Box 24-17, Survey of the University of British Columbia Faculty of Medicine, 11-13 January 1954.

35. Faculty of Medicine Fonds, Box 23-7, 16 May 1973, Flanagan to Bates; 24 August 1973, "An Analysis of the Review of the Department."

36. Faculty of Medicine Fonds, Box 27-8, 7 October 1974, Truman to Labour Relations Board; 4 October 1974, Margetts to Labour Relation Board.

37. Faculty of Medicine Fonds, Box 26-7, 19 November 1973, Polglase to Webber; Raines Collection, Job Offers file, 8 November 1974, Smith to MacNaughton and 15 January 1975, Smith to Friesen;

38. Faculty of Medicine Fonds, Box 16-1, [1978], notes on enrolment, expenditures, budgets, etc; Box 23-7, 17 June 1973, Bates to McCreary, Bates to Gage, Bates to Volkoff; Box 40-35, 14 July 1986, Smith to Webber. Community Relations Fonds, Box 7-21 President's Report 1977–78;

39. Smith Fonds, Box 1-19, 29 August 1973, Michael to Fred; 23 October 1973, Fred to Michael; 30 October 1973, Michael to Fred.

40. Personnel File, 16 December 1974, Mason to Smith and 23 December 1974, Polglase to Taylor; Smith Fonds, Box 1-19, 7 October 1974; 10 October 1974, Taylor to Smith; 17 October 1974, Smith to Taylor.

CHAPTER 4

1. Chapter 7 provides details of the Plus-Minus method.

2. Smith Fonds, Box 19-1 and 19-2, notebooks. Mike's Cambridge notebooks are the only notebooks that were available for this study.

3. Personnel File, 23 January 1976, Smith to Polglase. Hae, AluI, HpaIII, and HphI are the names of specific restriction endonuclease used to cut the DNA to prepare primers.

4. This program was one of the first for analyzing long DNA sequences. Duncan McCallum and Michael Smith, "Computer processing of DNA sequence data," *Journal of Molecular Biology* 116 (1977): 29–30.

5. F. Sanger, G.M. Air, B.G. Barrell, N.L. Brown, A.R. Coulson, J.C. Fiddes, C.A. Hutchison III, P.M. Slocombe, and M. Smith, "Nucleotide sequence of bacteriophage phiX174 DNA," *Nature* 265, no. 5596 (1977): 687–695.

6. Personnel File, 10 March 1977, Polglase to Bates; Smith Fonds, Box 1-19, 1 December 1976, Smith to Brown; Roger Lenn, "Virus violates an important biological law," *New Scientist* 21 (October 1976): 148. Despite the great

success, Sanger's lab soon adjusted the complex plus-minus method to the vastly simpler and more efficient "chain terminator" sequencing method. This method eventually found its use in the Human Genome Project.

7. Smith Fonds, Box 11-1, 2 February 1996, Smith to BCRI; Personnel File, 18 January 1976, Smith to Diane.

8. Personnel File, 6 October 1975, Smith to Polglase; 20 October 1975, Polglase to Smith; 18 January 1976, Smith to Diane.

9. C.A. Hutchison III and M.H. Edgell, "Genetic assay for small fragments of bacteriophage phiX174 deoxyribonucleic acid," *Journal of Virology* 8 (1971): 181–189.

10. Smith Fonds, Box 25-8, 28 August 1978, Kössel to Smith; 6 September 1978, Smith to Kössel; C. A. Hutchison III, S. Phillips, M.H. Edgell, S. Gillam, P. Jahnke, and M. Smith, "Mutagenesis at a Specific Position in a DNA Sequence," *Journal of Biological Chemistry* 253 (1978): 6551–6560.

11. The citric acid cycle plays a very important role in metabolism. Ingested carbohydrates are degraded to molecules that enter the citric acid cycle (also called the TCA or Krebs cycle) and are modified into intermediates that are "siphoned off" into important biosynthetic pathways including the synthesis of high-energy phosphate compounds.

12. Smith Fonds, Box 26-3, 8 May 1979, Smith to Soll; Box 26-12, 29 August 1980, Smith to Szbalski.

13. Smith Fonds, Box 1-14, Report to the Medical Research Council of Canada on research carried out during 1978–1983.

14. Smith Fonds, Box 24-13, 30 June 1977, Smith to Hamlyn.

15. Personnel File, 25 August 1978, Polglase to O'Brecht; Faculty of Medicine Fonds, Box 55-M12, 25 November 1988, Slotin to Webber; Smith Fonds, Box 1-14, 12 April 1979, Smith to Webber; 9 December 1983, Roxburgh to Smith; Box 1-15, Report for 1973–1978 on Smith Research.

16. Faculty of Medicine Fonds, Box 23-23, 4 May 1977, Copp Report; Box 23-24, 4 March 1976, McGeer to Kenny; 8 April 1976, newspaper clipping, "Fast diagnosis sought on UBC hospital"; 4 May 1976, Plan for Expansion of the University of British Columbia Medical School; 16 December, 1976, Newspaper clipping, "Medical school plans accepted."

17. Faculty of Medicine Fonds, Box 16-1, 13 December 1978, Polglase to Webber; [1978] Notes on enrolment , expenditures, budgets etc.; 11 October 1978, Polglase to Webber; Box 16-14, 22 March 1979, Polglase to Kenny; Box 40-35, 14 July 1986, Smith to Webber; Community Relations Fonds, Box 7-21, President's Report 1977–78 (Faculty of Medicine, Department of Biochemistry).

18. George E. Carter, *The Federal Impact of Financing Higher Education in Canada,* Occasional Paper 25 (Canberra: Centre for Research on Federal

Financial Relations of the Australian National University, 1982); Denise Michaud, *Federal Grants to Canadian Universities Prior to 1977: A Historical Perspective* (Ottawa: AUCC, 1982); Gordon Shrimpton, "A Decade of Restraint: The Economics of B.C. Universities," in *The New Reality: the Politics of Restraint in British Columbia,* eds. Warren Magnusson and others (Vancouver: New Star Books, 1984).

19. Community Relations Fonds, Box 7-21, President's Report 1977–78 (Faculty of Medicine, Department of Biochemistry); "Career virtually ended," *The Vancouver Sun* 18 March 1976, p. 5; Smith Fonds, Box 1-14, 11 April 1979, Smith to O'Brecht.

20. Smith Fonds, Box 1-14, 26 October 1978, Tener to Webber; 12 April 1979, Smith to Webber.

21. Smith Fonds, Box 1-5, 27 June 1977, Smith to Vogt; newspaper clipping [summer 1977]; "Congregation 1977," *UBC Reports,* 23 no. 9 (1977), p. 3; "Biochemist wins Biely prize," *UBC Reports* 23, no. 15 (1977), p. 1.

22. Smith Fonds, Box 1-5, 17 June 1981, Church to Smith; Box 1-5, 6 March 1981, Shrum to Smith; Box 26-12, 29 August 1980, Smith to Szbalski; Box 27-8, 25 September 1980, Setlow to Smith; Personnel File, 30 March 1981, Matheson to Smith.

23. Smith Fonds, Box 2-24, 26 June 1995, Transcript; Martin Kenney, *Biotechnology: The University-Industrial Complex* (New Haven: Yale University Press, 1986), 4, 28, chap. 1; Jane Sanders, *Into the Second Century* (Seattle: University of Washington Press, 1987), 35–42.

24. Faculty of Medicine, Box 19-13, 8 February 1982, Oles to Webber (Zymos/Smith contract); Smith Fonds, Box 11-1, CV of Ben Hall; Raines Collection, August 1987, ZymoGenetics Inc.–Company Background; "Unravelling the twisted threads of our genes," *The Vancouver Sun,* 3 January 1981, p. A10.

25. "Payoffs immense if science masters secrets of cells," *The Vancouver Sun,* 3 January 1981, p. A10.

26. Faculty of Medicine Fonds, Box 19-13, 22 June 1983 and 22 July 1983; Smith Fonds, Box 5-10, 5 May 1999, Smith to Kilburn.

27. Faculty of Medicine, Box 19-13, 8 February 1982, Oles to Webber (contract between Zymos and Smith); Raines Collection, ZymoGenetics file, 20 January 1986, Campos to Smith. Evidently the MRC had lifted its regulation prohibiting associates from accepting outside remuneration.

28. Smith Fonds, Box 1-19, 21 January 1981, Shirleen to Michael; Faculty of Medicine, Box 19-13, 10 February 1982, Smith to Bill.

29. Burton Feldman, *The Nobel Prize: A History of genius, controversy, and prestige* (New York: Arcade Publishing, 2000), 186.

30. Faculty of Medicine Fonds, Box 18-18, 19 January 1982, Dennis to

Bill; Box 20-10, 22 December 1983, Smith to Vance; Raines Collection, ZymoGenetics file, 13 April 1983, Senior Scientists to Founders.

31. Faculty of Medicine Fonds, Box 19-13, 26 January 1983, Price to Vance; Box 26-1, Report of Survey, 17–20 November 1980.

32. Faculty of Medicine Fonds, Box 20-10, 16 January 1984; 6 February 1984, Smith to Chalmers; 6 February 1984, Smith to Vance; 7 February 1984, Vance to Webber; Box 55-1, Biochemistry 400 Lecture Schedule, First Term 1984.

33. Faculty of Medicine Fonds, Box 19-13, 17 March 1983; Box 40-35, 8 December 1986, Bragg to Webber; Box 55-1, August 1985, Review of Biochemistry 400; Biochemistry 400 Student Evaluation Winter-Spring 1985.

34. Over 38,000 references to site-directed mutagenesis can be found today in PubMed, a database maintained by the National Library of Medicine. The database includes all peer-reviewed primary research publications in the life sciences. There are also more than 100,000 web hits for site-directed mutagenesis on the internet today.

35. Smith Fonds, Box 36-8, 7 May 1997, Ngsee to Smith; Box 10-7, 29 July 1994, Smith to Gilman.

36. "Innovative researchers win Science Council gold," *Vancouver Sun,* 23 October 1984; Smith Fonds, Box 1-5, 31 October 1984, Webber to Smith; Box 1-5, December 1984, Pack to Smith; [28 November 1984], Thompson to Smith.

37. Smith Fonds, Box 1-5, 9 June 1986, Hrica to Smith; 17 June 1986, Dirks to Smith; 1 July 1986, Smith to Hollenberg; 8 July 1986, Bushell to Smith.

38. "Canadians of Nobel calibre on rise," *The Globe and Mail,* 20 January 1987, p. A7; Faculty of Medicine Fonds, Box 40-35, 26 June 1986, Smith to Webber; Smith Fonds, Box 1-5, 20 November 1986, Smith to Strangway.

39. Smith Fonds, Box 1-5, Smith to Hrica and 29 October 1986, Hrica to Smith.

40. Faculty of Medicine Fonds, Box 40-35, 26 June 1986 and 11 July 1986, Smith to Webber.

CHAPTER 5

1. Faculty of Medicine Fonds, Box 40-22, 19 July 1982; Box 20-10, 12 December 1983, Vance to Webber; Box 52, Biotechnology Laboratory Report for 1988.

2. Crawford Kilian, *School Wars: The Assault on B.C. Education* (Vancouver: New Star Books, 1985); UBC Senate Fonds, 17 February 1982, Report of Senate Budget Committee.

3. "McGeer wants five-year outlook," *Ubyssey* (Vancouver), 29 January 1985, p. 3.

4. Faculty of Medicine Fonds, Box 18-18, 19 January 1982, Dennis to Bill; Box 21-12, 10 April 1985, Smith to Webber.

5. Smith Fonds, Box 1-5, 14 November 1986, Smith to Fraser; 11 June 1987, Smith to Campbell.

6. Personnel File, 9 March 1981, grant application; Smith Fonds, Box 6-10, 15 April 1989, Minutes; Faculty of Medicine Fonds, Box 20-10, 22 December 1983, Smith to Vance; Box 21-12, 8 May 1985, Michael to Bill; Box 40-36, 15 September 1986, MacGillivray to Webber; UBC Senate Fonds, 19 May 1982, Minutes (proposal).

7. Faculty of Medicine Fonds, Box 40-36, 25 August 1986, Mustard to Webber; 31 October 1985, Webber to Fredrickson.

8. Scrapbook File, 8 September 1988, Bragg to Charpenet (in re. MRC review); Faculty of Medicine Fonds, Box 35-33, 28 November 1985, Bois to Larkin; Faculty of Medicine Fonds, Box 40-35, 14 July 1986, Smith to Webber.

9. Faculty of Medicine Fonds, Box 40-35, 14 July 1986, Smith to Webber and 9 December 1986, Bragg to Birch; Box 40-36, 25 November 1986, Bragg to Zea.

10. Sheila Slaughter and Larry L. Leslie, *Academic Capitalism* (Baltimore: The Johns Hopkins University Press, 1997); UBC Senate Fonds, 17 February 1982, Report of Senate Budget Committee; Michael Bliss, *Beyond the Granting Agency: The Medical Research Council in the 1990s* (Ottawa: The Council, 2000).

11. Science Council of Canada, *Seeds of Renewal: Biotechnology and Canada's Resource Industries* (Ottawa: The Council, 1985); *Completing the Bridge to the 90s* (Ottawa: Natural Sciences and Engineering Research Council, 1985); Science Council of British Columbia, *A Research and Development Policy for British Columbia in the Eighties* (The Council: 1982); Science Council of Canada, *Biotechnology in Canada: Promises and Concerns* (Ottawa: The Council, 1981).

12. David Suzuki, "Biotechnology: Implications and Potential Biological Pitfalls" in Science Council of Canada, *Biotechnology in Canada: Promises and Concerns,* 11–21 (Ottawa: The Council, 1981); H.S. Ryan, "A Statement of Concern for Biotechnology" in Science Council of Canada, *Biotechnology in Canada: Promises and Concerns,* 55–62 (Ottawa: The Council, 1981); "Or have we been seduced by the promise of technology?" *The Vancouver Sun,* 3 January 1981, p. A10; BL Files, "Administrative Structure of the Molecular Genetics Centre" (Final version).

13. Raines Collection, Job Offers file, 8 August 1986, Smith to Richmond.

14. Smith Fonds, Box 33-2, [1992], "Remarks for 8th September Lunch with Dr. D.W. Strangway".

15. Faculty of Medicine Fonds, Box 52, Biotechnology Laboratory Report 1988 (Draft).

16. *UBC: Engine of Recovery* (Vancouver: The University of British Columbia, 1986); David Strangway, *Second to None* (Vancouver: The University of British Columbia, 1989); "Over the top," *University Affairs* 32 (2), p. 19; *University of British Columbia Fact Book* (Vancouver: Office of Budget and Planning, 1991).

17. BL Files, 16 May 1986, Detailed Biotechnology Proposal-Centres of Excellence.

18. Faculty of Medicine Fonds, Box 46-21, 6 January 1987, Goard to Ivan/Birch/Fraser; Smith Fonds, Box 1-5, 17 June 1986, Minutes, Faculty of Medicine Executive Meeting; BL Files, 16 May 1986, Detailed Biotechnology Proposal-Centres of Excellence.

19. BL Files, 6 November 1986, Minutes, Biotechnology Laboratory Review Committee; Smith Fonds, Box 1-17, 1 December 1986, Birch to Michael.

20. Faculty of Medicine Fonds, Box 20-10, 22 December 1983, Smith to Vance; Smith Fonds, Box 1-17, 24 April 1987, Birch to Dybikowski; Smith Fonds, Box 1-5, 20 November 1986, Smith to Strangway; Raines Collection, ZymoGenetics file, ZymoGenetics Consulting Agreement, 29 January 1987.

21. BL Files, December 1986, notes; Faculty of Medicine, Box 40-35, 8 December 1986, Bragg to Webber; Box 43-19, 24 April 1987, Birch to Webber; Box 58-Ad.5-3, 15 April 1987, Webber to Bragg; Smith Fonds, Box 1-17, 23 March 1987, Bois to Webber.

22. Smith Fonds, Box 1-5, 1 December 1986, Birch to Michael; BL Files, 11 February 1987.

23. Faculty of Medicine, Box 46-21, 5 February 1987, Report of the Director, UBC Biotechnology Laboratory; Box 57-3, [1989], Smith to President's Advisory Committee; BL Files, 1992–93 Budget, 1993–94 Budget.

24. BL Files, 6 November 1986, Minutes, Biotechnology Laboratory Review Committee; Faculty of Medicine Fonds, Box 48-5, 15 December 87, Strangway to Hagen; Box 48-34, 13 April 1987, "British Columbia Universities and the New Economy" (draft); 8 July 1987, "University of British Columbia: A Special Campaign."

25. Faculty of Medicine Fonds, Box 46-21, 28 January 1987, Smith to Beaumont; Box 52 Biotechnology Laboratory Annual Report, 1987–88 (Draft).

26. Faculty of Medicine Fonds, Box 46-21, 16 June 1987, Wilson to Haller; 19 August 1987, McBride to Smith; 3 February 1987, Birch to Deans

et al; BL Files, 1 May 1989, Minutes.

27. BL Files, 6 November 1986, Minutes, Biotechnology Laboratory Review Committee; 24 February 1987; 5 June 1987, Minutes; 8 June 1987, Guidelines for Funding by the Biotechnology Laboratory; Faculty of Medicine Fonds, Box 43-23, 21 February 1987, Smith to Dirks; Box 46-21, 11 February 1987, 12 March 1987.

28. BL Files, 6 May 1987, Minutes; Faculty of Medicine Fonds, Box 48-34, 10 March 1987, Arbitration Award; 24 April 1987, Jones to Fibiger; 8 July 1987, Ruling of the Commissioner; Box 57-3, [1989], Smith to President's Advisory Committee; Box 58-Ad5-3, 15 March 1989, Bragg to Ellison; 17 May 1989, Bragg to Sadowski.

29. BL Files, December 1986, notes; 28 January [1987] "Things to discuss with D. Birch"; 7 December 1987, Smith to Advisory Committee; 1992–93 Budget, 1993–94 Budget; Faculty of Medicine, Box 52-2-1, 27 May 1988, Flores to Webber.

30. BL Files, 8 April 1988, Agenda. The first professorial appointments included Brett Finlay, Wilf Jeffries, Terrance Snutch, James Kronstad, John Carlson, Louise Glass, Robin Turner, James Piret, and, briefly, James Chamberlain. Loida Escote-Carlson became first supervisor of the teaching laboratory; BL Files, [mid-1989], "Meeting with Dr. Birch: Questions.

31. BL Files, 16 January 1990, Minutes (Director's Report); Faculty of Medicine Fonds, Box 62-16, Biotechnology Laboratory Annual Report, 31 March 1990.

32. BL Files, 5 June 1987, Minutes; BL Files, 8 June 1987, "Guidelines for Funding"; Faculty of Medicine Fonds, Box 57-3, 15 December 1989, Smith to Miller; Smith Fonds, Box 1-17, 1 December 1986, Birch to Michael.

33. Faculty of Medicine Fonds, Box 57-3, 15 December 1999, Smith to Miller; Smith Fonds, Box 6-10, 1 August 1989, "Mapping and sequencing the human genome: a program for Canada."

34. BL Files, 20 September 1990, Smith to Birch.

35. Faculty of Medicine Fonds, Box 51-6, 21 June 1988, Braley to Smith; 30 June 1988, Webber to Smith; 12 September 1988, Braley to Smith; Smith Fonds, Box 5-10, 5 May 1999, Smith to Kilburn and 17 May 1999, Kilburn to Cairns; Raines Collection, ZymoGenetics file, "ZymoGenetics Inc. Company Background, August 1987" and "Offer to Purchase All Outstanding Shares of Capital Stock . . . ."

36. Raines Collection, ZymoGenetics file, "Offer to Purchase All Outstanding Shares of Capital Stock. . . ."

37. Smith Fonds, Box 1-5, 30 January 1989, Smith to Smith; 1 February 1989, Smith to Smith.

38. Smith Fonds, Box 1-5, 16 June 1989, Diana to Mike.

39. Janet Atkinson-Grosjean, "Adventures in the Nature of Trade: The Quest for 'Relevance' and 'Excellence' in Canadian Science" (unpublished PhD thesis, the University of British Columbia, 2001), 76; Smith Fonds, Box 6-10, 16 November 1995, Unruh to Smith.

40. Faculty of Medicine Fonds, Box 40-36, 25 August 1986, Mustard to Webber; Smith Fonds, Box 6-10, 16 November 1995, Unruh to Smith; Box 6-10, 1 August 1989, "Mapping and sequencing the human genome: a program for Canada."

41. Faculty of Medicine Fonds, Box 51-N8, 13 January 1988, Weiler to Bragg/McBride/Schrader/Smith.

42. Faculty of Medicine Fonds, Box 52-2 (U2-1-2), 30 June 1988, Memo.

43. "Centres of Excellence Network in Protein Engineering: 3-D Structure, Function and Design," PENCE Final Proposal, 4 October 1990; Author's personal collection, "Position Paper from Allelix, Inc. MOSST Conference — Computer Aided Design of Proteins," 24–25 March, 1988.

44. BL Files, 1 December 1989, Miller to Smith; PENCE Final Proposal, 1.5iii; Faculty of Medicine Fonds, Box 57-B.4, 14 August 1987, Weiler to Bragg; 26 August 1987, Smith to Weiler; 3 November 1989, Smith to Bragg; 6 December 1989, Smith to Bragg/Weiler.

45. Faculty of Medicine Fonds, Box 57-3, Biotechnology Laboratory Faculty Research Support; Box 62-16(B.4), Biotechnology Annual Report 1989–90.

46. Scrapbook File, 8 September 1988, Bragg to Charpenet; Smith Fonds, Box 1-14, Report to the MRC on research carried out 1983–1988; 28 March 1989, Slotin to Smith.

47. Raines Collection, Arnold School file, 4 September 1985, Smith to Rhodes; 16 September 1985, Rhodes to Smith; Smith Fonds, Box 2-1, 5 February 1988, Transcript.

48. "UBC Scientist wins prize for genetics work," *The Globe and Mail,* 24 August 1988, p. A8; Smith Fonds, Box 1-5, 24 August 1988, Kennedy to Smith; Scrapbook File, 23 December 1988, Tener/Bragg to Keough.

49. Faculty of Medicine Fonds, Box 48-19, 19 November 1987, Smith to Owen; Smith Fonds, Box 2-1, 5 February 1988, Convocation address; Box 62-16 (B.4), Biotechnology Laboratory Annual Report, 31 March 1990; Smith Fonds, Box 1-4a, 14 November 1991, Smith to Miller.

50. Smith Fonds, Box 7-11, 8 July 1987, Smith to Strangway; 8 April 1988, Smith to McInnes; 27 February 1991, Smith to Masse; Author's Personal Collection, 11 January 1987, Smith to Rubin and Smith to Mauk.

51. Giles Julien, "The Funding of University Research in Canada Current Trends," *Higher Education Management* 1, 1 (March 1989): 66–72; Faculty of Medicine Fonds, Box 62-B.4, Biotechnology Laboratory Annual Report,

1989–90; BL Files, Lists of grants, 1992–1995; Smith Fonds, Box 36-4, 13 October and 17 October 1989, Smith to Miller.

52. Faculty of Medicine Fonds, Box 52-2, Biotechnology Laboratory Annual Report 1987/1988; Smith Fonds, Box 5-10, 8 March 1999, Petter to Piper.

53. Faculty of Medicine Fonds, Box 62-16 (B.4), 2 March 1990, Tan to Smith and 2 March 1990, Smith to Strangway; Smith Fonds, Box 4-5, 11 January 1999, Smith to Tan and 9 February 1999, Tan to Smith.

54. Faculty of Medicine Fonds, Box 48-34, 13 April 1987, "British Columbia Universities and the New Economy (draft)"; 8 July 1987, "University of British Columbia: A Special Campaign." James. L. Turk, ed., *The Corporate Campus* (Toronto: James Lorimer & Company, 2000); David Suzuki and Holly Dressel, *From Naked Ape to Superspecies* (Toronto: Stoddard, 1999).

55. BL Files, 11 Sept 1990, Birch to PACB; BL Files, [mid-1989], "Meeting with Dr. Birch: Questions; 18 September 1992, McBride to Webber.

56. Faculty of Medicine Fonds, Box 63-Ad. 5-3 [1990], Department of Biochemistry Resources and Achievement Profile; Author's Personal Collection, 11 February 1991, Smith to Faculty.

57. BL Files, 9 April 1991, Minutes, Biotechnology Steering Committee; Smith Fonds, Box 5-9, 1 May 1991, Draft of employment contract; 1 May 1992, Smith to Perry; Box 1-5, 19 June 1992, Smith to Webber.

58. Faculty of Medicine Fonds, Box 47-17, 11 September 1987, Schrader to Webber.

59. "Report warned Fox foundation faced difficulties," *The Vancouver Sun,* 4 September 1991, p. B3; "High marks couldn't save scientist's job," *The Vancouver Sun,* 21 August 1991, p. B3; "Six senior scientists say they feared for future of biomedical centre," *The Vancouver Sun,* 23 August 1991, p. B7; "NDP demands investigation of Fox foundation," *The Vancouver Sun,* 22 August 1991, p. A13; "Fox fiasco shows governments still misspending our money," *The Vancouver Sun,* 24 August 1991, p. B9; "Perry charges Socreds covering up damaging report on Fox Foundation," *The Vancouver Sun,* 18 September 1991, p. B5.

60. "Terry Fox Foundation Trustees defend use of public funds, trust," *The Vancouver Sun,* 28 August 1991, p. B7.

61. "Politics, conflict plague organization, probe concludes," *The Vancouver Sun,* 24 October 1991, p. E7; "Fox foundation counterpunches financial critics," *The Vancouver Sun,* 30 October 1991, p. B5; Smith Fonds, Box 5-9, 27 September 1991, Smith to Warren and Smith to Crumpton.

62. "Terry Fox Foundation Trustees defend use of public funds, trust," *The Vancouver Sun,* 28 August 1991, p. B7; "High marks couldn't save scientist's job," *The Vancouver Sun,* 21 August 1991, p. B3; "Six senior scientists say

they feared for future of biomedical centre," *The Vancouver Sun,* 23 August 1991, p. B7; "Instability threatens biomedical centre: Director's dismissal angers staff," *The Vancouver Sun,* 19 July 1995 (final edition), p. A1; Smith Fonds, Box 1-4a, 14 November 1991, Smith to Miller.

63. Smith Fonds, 23 September 1992, Smith to Lavkulich; Box 33-2, [1992], "Remarks for 8th September Lunch with Dr. D.W. Strangway"; author's personal collection, [December] 1992, Michael to (various).

64. Smith Fonds, Box 5-9, 29 November 1991, Smith to Crompton; "UBC rescues Fox research lab, *The Vancouver Sun,* 3 October 1991, p. B1; "Terry Fox research foundation will close," *The Vancouver Sun,* 14 March 1992, p. A1.

65. "Instability threatens biomedical centre: Director's dismissal angers staff," *The Vancouver Sun,* 19 July 1995 (final edition), p. A1. The rehired Director even created commercially valuable products and at least one very successful spin-off company.

66. Smith Fonds, Box 5-9, 1 May 1992, Smith to Woodward and Smith to Perry; Box 1-5, 5 June 1992, Woodward to Smith; 19 June 1992, Smith to Webber.

67. Smith Fonds, Box 5-9, 1 May 1991, Draft of employment contract; Smith Fonds Box 33-2, 10 June 1991, Smith to Warren; "Remarks for 8th September lunch;" 2 September 1997, Smith to Withers.

68. Smith Fonds, Box 1-4a, 14 November 1991, Smith to Miller; BL Files, 16 March 1992, Minutes; BL Files, 16 March 1992, Minutes, PSCB; [March 1992], McBride to Birch; BL Files, 20 September 1990, Smith to McBride.

69. BL Files, 21 December 1992, 11 Jan 1993, Birch to various; Teaching reviews were reaffirmed in 1995. UBC Senate Minutes, 15 November 1995; BL Files, 8 February 1993, McBride to Birch.

70. BL Files, 26 October 1992, Watkinson to Chemical Engineering faculty; 2 April 1993, Smith to Binkley; 21 July 1993, Smith to Heads; 20 April 1993, Smith to McBride.

71. BL Files, 20 April 1993, Smith to McBride; 30 July 1993, Smith to McBride.

72. PENCE Second Annual Report, 31 March 1992, 125; Third Annual Report, 31 March 1993, 1.3, 1.5.

73. Smith Fonds, Box 33-2, 4 August 1992, Smith to Söll, and other papers.

CHAPTER 6

1. Smith Fonds, Box 2-12, 1994, Vancouver Institute Lecture (transcript); Interview, *As It Happens,* CBC Radio, 13 October 1993; "UBC's Nobel Prize prof feted," *The Vancouver Sun,* 14 October 1993, p. A1.

2. "Offbeat," *UBC Reports,* 28 October 1993, p. 3; "Nobel Smith gets early lesson in handling celebrity status," *The Vancouver Sun, 16 October 1993,* p. F12; "Nobel Prize winner finds pace is hectic before trip," *The Vancouver Sun,* 26 November 1993, p. A3.

3. "People," *UBC Reports,* 25 November 1993, p. 7.

4. "Offbeat," *UBC Reports,* 28 October 1993, p. 3; Raines Collection, Journal, 19 October 1993.

5. The economics award is actually the Central Bank of Sweden Prize in Economic Science in Memory of Alfred Nobel.

6. Department of Biochemistry, Scrapbook File, 24 January 1991, Bragg to Nobel Committee.

7. Smith Fonds, Box 10-4, 21 March 1994, Smith to Tustanoff.

8. "In a Nobel fashion," *The Vancouver Sun,* 30 November 1993, p. C1.

9. "Winner worried about funding," *The Vancouver Sun,* 14 October 1993, p. A2; "B.C. scientist awarded Nobel," *The Globe and Mail,* 14 October 1993, p. A1; "Nobel Prize winner finds pace is hectic before trip," *The Vancouver Sun,* 26 November 1993, p. A3.

10. Harry Black, *Canada and the Nobel Prize* (Markham: Pembroke, 2002). John Macleod, who shared the Nobel with Banting, was British and is often counted as a Canadian. "Nobelist Uses New Clout," *The Province,* 17 October 1993, p. A20.

11. Smith Fonds, Box 6-10, 15 April 1989, Minutes; Faculty of Medicine Fonds, Box 20-10, 22 December 1983, Smith to Vance.

12. Raines Collection, Schizophrenia file, various letters.

13. "Nobel Prize winner donates $500,000," *UBC Reports* 9 December 1993, p. 1; "Nobel winner tired but proud," *The Vancouver Sun,* 21 December 1993.

14. Clifford Thompson, ed., *Nobel Prize Winners — 1992–1996 Supplement* (New York: H.W. Wilson, 1997), 99–102; www.barbarahendricks.com.

15. CBC Newsworld, 10 December 1993; "Muffins with Michael mark Nobel day," *The Vancouver Sun,* 11 December 1993, p. A1.

16. *Maclean's,* 24 January 1994, p. 22; Author's Personal Collection, 4 February 1994, Smith to Mauk.

17. Smith Fonds, Box 4-1, 28 March 1995, Smith to Collins; Box 14-22, Report of the BC/Taiwan Biotechnology R&D Mission; "Our Nobel winner spends a year on the road," *The Vancouver Sun,* 1 October 1994, p. A3; "Nobel winner's life just a hullabaloo," *The Vancouver Sun,* 27 January 1995, p. B1.

18. Smith Fonds, Box 1-13, 13 April 1994, Strangway to Smith; 13 September 1994, Birch to Webber; Box 1-18, 28 November 1995, Birch to Michael.

19. Smith Fonds, Box 1-13, 13 September 1994, Birch to Webber; 8 No-

vember 1994, Birch to Smith; "Nobel winner relied on Scholarship for education," *The Ubyssey,* 19 October 1993, p. 1; "And they earn every penny: the well paid men of UBC," *The Ubyssey,* 15 March 1994, p. 1; "Students screwed, UBC pigs on top," *The Ubyssey,* 25 March 1994, p. 1.

20. Smith Fonds, Box 1-18, 3 October 1997, Michael Smith's Activities, 1996; Box 4-1, 14 February 1996, Smith to Reece; Transcripts in Smith Fonds, Boxes Box 2-7, 2-22, 2-23, 2-24, 2-33, 2-26, 2-45, 2-28; 2-32; "Convocation Address," *Canadian Society for Biochemistry and Molecular & Cellular Biology Bulletin* (1996): 33–35.

21. Smith Fonds, Box 4-1, 26 February 1996, Fawcett to Smith; 6 March 1996, Smith to Fawcett; Box 7-3, 15 December 1994, Smith to Johnson.

22. Smith Fonds, Box 9-14, 12 April 1995; Box 9-16, various letters; Box 7-12, 7 July 1997, Union of Concerned Scientists to Smith; Box 7-16, 3 February 1997, Smith to Chrétien; Smith Fonds, Box 4-1, 9 January 1996, Smith to Field; Box 7-9, 2 June 1997, Watson to Smith; Box 13-8, 10 October 1997, Annan to Smith.

23. Smith Fonds, Box 7-12, 8 March 1996, Smith to Suzuki; Box 7-18, 16 September 1998, Smith to Lawrence; Box 4-6, 25 October 1999, Smith to Daynard; Smith Fonds, Box 16-5, 9 February 1999, Smith to Hackl.

24. Smith Fonds, Box 7-12, 24 June 1997, Smith to Suzuki; Box 7-16, 23 October 1996, Smallwood to Smith; 24 June 1997, Parras to Scudder.

25. Smith Fonds, Box 4-5, 15 June 1999, Herbert to Smith; 24 June 1999, Smith to Herbert; Box 7-13, 15 March 2000 Smith to Pugwash; Box 13-2, 24 May 2000, Smith to Watkins; Box 13-5, various letters and newsletters, including 19 May 1999, Smith to Eigendorf.

26. Smith Fonds, Box 4-4, 6 July 1998, Smith to Schaffer; Box 6-4, 12 November 1999, Smith to Atkinson.

27. Smith Fonds, Box 7-15, 12 July 1999, Smith to Alain; 8 November 1999, Smith to Alain; Box 4-6, 25 October 1999, Smith to Daynard; Box 4-8, 24 January 2000, Smith to Knelman; 9 April 2000, Smith to Duff; Box 7-12, 24 May 2000, Smith to Batycki; 8 August 2000, Suzuki to Smith.

28. BL file, 13 February 1995, Smith to McBride; 10 April 1995, Minutes.

29. Scrapbook File, 31 August 1994, Smith to Slotin.

30. "Crisis Point focus for Institute study," *UBC Reports,* 4 April 1996, p.4; "Wall Institute to meld best minds in research," *UBC Reports,* 12 December 1996; "Institute's tack questioned," *UBC Reports,* 23 January 1997. Smith Fonds, Box 36-2, 29 November 1996, MacCrimmon to Wall Associates; 11 February 1998, Minutes.

31. Smith Fonds, Box 7-17, 7 February 1997, Smith to Granot; 23 April 1997, Memo From President; Box 9-18, 20 January 1998, Smith to Piper.

32. Smith Fonds, Box 1-22, 1 May 1995, Jun to Thompson, Jun to Ching;

16 August 1995, Katz to Smith; Box 33-2, 18 March 1998, Khorana to Smith; Box 1-21, 18 May 1994, Smith to Slotin; 9 April 1996, Smith to Sniekus; 31 October 1996, News release.

33. Smith Fonds, Box 1-23, 23 July 1997, Triggle to Walker; 30 June 1995, Hakim to Smith; Box 1-24, 21 June 1996, Bell to Smith; [n.d.] list of awards. Ciba Geigy later became Novartis after merging with Sandoz Canada.

34. Smith Fonds, Box 8-4, 26 June 1998, Smith to Dirks; Smith Fonds, Box 7-13, various letters, memos, reports; Smith Fonds, Box 1-1, CV; Box 1-22, 6 January 1996, Smith to Rix.

35. "A Conversation With Dr. Michael Smith," *CFBS Newsletter* 12, 1 (Spring 1994): 11; "Nobel Canada," *Innovation,* 4 (June 1995): 5–12; Smith Fonds, Box 3-11, 27 October 1998, Transcript; Box 4-7, 9 November 1999, Smith to Lemieux; Box 6-4, 13 August 1998, Smith to Atkinson; Box 6-12, 13 June 1999, notes for lunch meeting speech.

36. Smith Fonds, Box 9-10, 27 April 1995, Smith to Wosnick; Box 5-1, 4 March 1997, Smith to Bruneau (CAUT brief, 1997); Box 15-4, 20 December 1994, Smith to Martin; "Nobel winner's life just a hullabaloo," *The Vancouver Sun,* 27 January 1995, p. B1.

37. Smith Fonds, Box 12-7, 5 May 1997, Smith to McBride; *UBC Reports,* 19 September 1996, p. 19.

38. Smith Fonds, Box 5-1, 28 January 1997, Smith to Martin; 7 July 1997, Drapeau to Smith; 18 July 1997, Smith to Drapeau; Box 8-19, 8 March 2000, Lynch to Smith; Smith Fonds, Box 2-52, 31 October 1996, "Comments to Prime Minister Jean Chrétien at meeting of Advisory Council."

39. "Research money praised: main concern is possible neglect of pure research," *Kingston Whig Standard,* 20 February 1997, p. 3; Smith Fonds, Box 5-1, 22 April 1997, Smith to Martin; Box 5-12, Communiqué, 19 October 1998; Box 5-1, 10 October 1997, Smith to Martin.

40. Smith Fonds, Box 4-7, 1 November 1999, Cochrane to Smith; Box 6-13, 14 December 1994, Stiller to Smith; 8 March 1996, Stiller to Martin; Box 6-14, 25 September 1996, Smith to Stiller; Box 6-16, Announcement about the Friesen-Rygiel Prize.

41. Smith Fonds, Box 7-18, 16 November 1998, Smith to Rock; Box 6-4, 13 August 1998, Smith to Atkinson; Box 6-12, 17 June [1997], "Should the Canadian Institutes of Health Research Have a Commercialization Mandate"; Box 9-7, 1 May 1997, Smith to Friesen; 6 April 1998, Proposal; 8 July 1998; 15 October 1998, Smith to Friesen; Smith Fonds, Box 4-1, 22 November 1996, Smith to Polanyi.

42. Neil Tudiver, *Universities for Sale: Resisting Corporate Control over Higher Education* (Toronto: James Lorimer, 1999); Pat Armstrong and Hugh Armstrong, *Wasting Away* (Toronto: Oxford University Press, 1996).

43. Smith Fonds, Box 4-1, 13 February 1996, Smith to Tang; 20 June

1996, Smith to Eltis; 31 July 1996, Smith to Friesen; Box 15-9, 17 June 1996, Schedule of events; Box 9-6, 17 January 1996, Smith to Polglase. Huyer received her PhD posthumously, without a defence.

44. Smith Fonds, Box 1-20, 13 March 1996, Barry to Michael; Box 9-7, 1 May 1997, Smith to Friesen; Box 1-14, 9 January 1996, Friesen to Hollenberg.

45. Smith Fonds, Box 9-7, 1 May 1997, Smith to Friesen; Box 1-20, 22 April 1997, Miller to Bressler.

46. The final count will be certainly less than 100,000.

47. Smith Fonds, Box 4-1, 11 April 1996, Friesen to Scriver; Box 6-8, 25 February 1993, Gray to Smith; Box 6-10, 1 August 1989, "Mapping and sequencing the human genome"; 17 August 1989, Doolittle to Colleagues; 5 September 1990, Mustard to Winegard; Box 9-2, 28 May 1990, Worton to Bois; 19 March 1991, Smith to Scriver; Smith Fonds, Box 4-1, 26 March 1996, Scriver to Freisen; 11 April 96, Friesen to Scriver; 26 April 1996, Scriver to Smith; Abby Lippman, Karen Messing, Francine Mayer, "Is Genome Mapping the Way to Improve Canadians' Health?" *Canadian Journal of Public Health* 81 (1990): 397–8.

48. Smith Fonds, Box 5-2, 4 March 1996 and 22 December 1996; Box 6-5, 10 July 1996, Altow to Smith; "B.C. Cancer Agency sets fundraising target at $100 million," *The Vancouver Sun,* 8 April 1997.

49. "3 Tenors gala to benefit research," *The Vancouver Sun,* 23 November 1995, p. B1 (final edition); "Vancouver gets gene decoding centre," *The Vancouver Sun,* 8 October 1997, p. B5; Smith Fonds, Box 11-5, 2 December 1997, Smith to Eltis.

50. Smith Fonds, Box 9-7, 1 May 1997, Smith to Friesen; Box 5-3, 16 April 1997, Hugerford to Smith; 9 September 1997, Matthews to Paris; Box 4-4, 6 July 1998, Smith to Schaffer; "The Art of Discovery," *The Vancouver Sun,* 9 May 1998.

51. "Vancouver gets gene decoding centre," *The Vancouver Sun,* 8 October 1997, p. B5.

52. Smith Fonds, Box 1-4, n.d. Strangway to Smith; Box 5-12, 14 July 1997, Mackie to Faculty; 8 January 1998, Smith to Evans; 30 March 1998, Evans to Smith; 11 August 1998, Smith to Strangway. Mike excused himself from voting on another occasion.

53. Smith Fonds, Box 4-9, 28 July 2000, Smith to Slaymaker; Box 5-12, 17 March 1998, Gagnon to Board; 27 April 1998, Speer to Smith.

54. Smith Fonds, Box 1-9, 6 December 1999, Smith to Cleghorn; Box 4-4, 10 September 1998, Smith to Gadey; 15 December 1998, Smith to Garder; Box 4-5, 15 March 1999, Smith to [Davies]; Box 6-16, 3 December 1998, Smith to Stiller; Box 9-1, 2 December 1999, Lapierre to Smith.

55. "Health research needs funds: Smith," *UBC Reports,* 15 October 1998,

p. 7; Smith Fonds, Box 7-18, 22 March 1999, Smith to Adams; 11 April 1999, Adams to Smith. Mike's Member of Parliament from Vancouver, Ted McWhinney, was also a member of that committee.

56. Smith Fonds, Box 4-5, 15 March 1999, Smith to [Davies]; Box 5-10, 29 July 1999, Mitchel to e-mail list; Box 5-13, [19 May 1999], CIG Proposal to CFI; [4 October 1999], Ling to UBC/CFI Committee Members; Box 8-15, "The Genome Sequence Centre"; "Mapping the genetic highway," *Maclean's* (26 July 1999): 46–47.

57. Smith Fonds, Box 5-4, 15 October 1999, Smith to Miller; 21 February 2000, McFarlane to Smith; Box 5-5, 24 June 1999, Whitney to Smith; 30 June 1999, Smith to Whitney.

58. Smith Fonds, Box 8-13, 10 May 2000, Smith to LePage; 11 May 2000, Smith to Foxall; 3 August 2000, "Intellectual Property Policy" (draft).

59. Smith Fonds, Box 8-15, "The Genome Sequence Centre"; 17 July 2000, News Release; "UBC research labs get $35 million boost," *The Vancouver Sun* [final edition], 13 June 2000, p. A3; "B.C. Cancer Agency nears goal with bold fund-raising move," *The Vancouver Sun* [final edition], 29 November 2000, p. B6.

60. Smith Fonds, Box 1-10, list; Box 4-2, 22 December 1996; Box 9-8, 27 May 1996, Smith to Khorana; John Bishop and Dennis Green, *Simply Bishops* (Vancouver: Douglas and McIntyre, 2002), dedication; "UBC Nobel scientist a double winner," *The Vancouver Sun,* 2 June 1999, p. A10.

61. Smith Fonds, Box 7-18, 13 March 2000, Smith to Chrétien; Carl Berger, *Honour and the Search for Influence: A History of the Royal Society of Canada* (Toronto: University of Toronto Press, 1996).

CHAPTER 7

1. DNA is the genetic material in most organisms and many viruses. Some viruses, however, have an RNA genome. These include the HIV virus, poliovirus, and the recently discovered SARS coronavirus.

2. Albert Lehninger, *Principles of Biochemistry,* 3rd edition, eds D. L. Nelson and M.M. Cox (New York: Worth Publishers, 2000).

3. J.D. Watson and F.H. Crick, "Molecular structure of nucleic acids; a structure for deoxyribose nucleic acid," *Nature* 171, no. 4356 (April, 1953): 737–738.

4. There are 64 triplet combinations: 4 bases raised to the power 3. Three are stop codons hence only 61 specify a particular amino acid. Since there are only 20 amino acids, some amino acids are coded by multiple triplets. For example, phenylalanine has four codons: UUU, UUA,UUC, and UUG.

5. This abbreviation for the trimer indicates that there is an A, C, and T

nucleotide coupled together via phosphodiester bonds represented by the lower case "p." A further abbreviation often used is pACT in which the "p" representing the internal phosphodiester bonds is omitted. The d indicates that the oligomer is a deoxyribonucleotide.

6. One problem with this type of chemical synthesis is that some people are remarkably sensitive to DCC, which is moderately volatile. Even weighing out the reagent in a chemical fume hood and carrying out all of the steps in the fume hood, those sensitive to DCC eventually develop a very itchy rash all over their bodies. Although Mike had used this chemical for years without problems, Caroline soon became highly sensitive to it as did a summer student who developed the rash without ever handling the chemical. The chemical fume hoods used in the 1960s were not nearly as efficient compared with those in use today.

7. The water soluble carbodiimide was not volatile hence this chemical did not elicit a rash.

8. DNA is a very long molecule which is readily broken up into fragments of ~50,000 base pairs simply by pipetting or stirring mechanically in solution.

9. C. R. Astell, "Thermal Elution of Oligonucleotides on Cellulose Columns Containing Oligodeoxyribonucleotides of Defined Length and Sequence," (unpublished PhD thesis, the University of British Columbia, 1970).

10. I. Gillam, S. Millward, D. Blew, M. von Tigerstrom, E. Wimmer, and G.M. Tener, "The separation of soluble ribonucleic acids on benzoylated diethylaminoethylcellulose," *Biochemistry* 6 (1967): 3043–3056.

11. S. Gillam, P.A. Jahnke, and M. Smith, "Enzymatic synthesis of oligodeoxyribonucleotides of defined sequence," *Journal of Biological Chemistry* 253 (1978): 2533–2539.

12. The procedure begins with a single-stranded DNA template (i.e. the genome of phiX174 virus). A short piece of the phiX174 DNA was obtained by cutting the double-stranded form of the viral DNA with a particular restriction endonuclease and isolating a specific fragment. The specific fragment was heated to separate the two strands and allowed to bind to the template strand according to matching AT and GC base pairs. An enzyme, DNA polymerase, extends the oligomer in the same way cells are able to copy their DNA during DNA replication. The substrates for this reaction include all four of the bases found in DNA (in the form of nucleoside triphosphates), one of which is labeled with P^{32}. This extension step is random and the extended fragments vary in length according to the number of bases added. The template and extended oligomer complexes are then purified to remove unincorporated nucleotides and divided into eight small tubes, four for the "plus" reactions and four for the "minus" reactions.

Each of the "plus" tubes are incubated with a bacterial virus (T4) DNA polymerase with one of the triphosphates of A, T, C, or G (hence the name of each tube: +A, +T, +C, and +G). This T4 DNA polymerase "chews back" (degrades) the extended oligomers until it reaches a base that corresponds with the nucleotide added to that tube. Hence oligomers in tube +A all end with A, those in tube +T all end with a T, and so on.

The "minus" tubes (-A, -T, -C, and –G) are incubated with a different DNA polymerase isolated from *E. coli* and three of the four deoxynucleoside triphosphates, the substrates for DNA synthesis. For example, in the –A tube the triphosphates of T, C, and G are inserted by the polymerase as specified by the template strand, but the reaction stops when it needs an A. Similarly oligomers in the -T, -C, and -G tubes are stopped when a T, C, or G are specified by the template strand.

Oligomers from both the "plus" and "minus" reactions are separated from the template strand and resolved according to size by gel electrophoresis. The gel containing the radioactive extended oligomers from both the "plus" and "minus" reactions is dried and exposed to X-ray film to detect the radioactive oligomers. The scientist then reads the gel pattern from bottom to top to interpret the order of the bases. Although this method was used to determine the first genome sequenced enzymatically, that of phiX174, the method was complex and the results were not always unambiguous. For example, cumulative runs of a single base were difficult to read.

13. F. Sanger, G.M. Air, B.G. Barrell, N.L. Brown, A.R. Coulson, J.C. Fiddes, C.A. Hutchison III, P.M. Slocombe, and M. Smith, "Nucleotide sequence of bacteriophage phiX174 DNA," *Nature* 265, no. 5596 (1977): 687–695.

14. In the chain terminator method to determine the sequence of the bases in DNA, each tube contained a template (e.g. phiX174 DNA), a short oligomer (restriction fragment) which bound to the template, all four dideoxynucleoside triphosphates (A, T, C, and G), and a dideoxynucleoside triphosphate of either A, T, C, or G (ddA, ddT, ddC, or ddG). The addition of ddA by DNA polymerase to the elongating strand in the A tube prevented the addition of more nucleotides and thus the A tube contained a series of DNA fragments terminated with a ddA. The same held for other tubes and ddT, ddC, and ddG. One could readily sequence up to three hundred bases by analyzing the bands found in the tubes. This chain-terminator method proved to be very robust and was further modified to use dideoxynucleotides of A, T, C, and G coupled with a different coloured fluorescent tag. This allowed all four reactions to take place in the same tube and analysis of the products in a single gel lane or capillary. Further improvements in separating the fragments allowed the determination of stretches of DNA as long as 1,000 bases with each reaction. This procedure was automated and used very successfully in the sequencing of the human genome.

15. E.S. Lander, et al., "Initial Sequencing of the human genome," *Nature* 409 (2001): 860–921. The "et al." includes over 100 authors!

16. C.A. Hutchison III and M.H. Edgell, "Genetic assay for small fragments of bacteriophage phiX174 deoxyribonucleic acid," *Journal of Virology* 8 (1971):181–189.

17. The Genetic Code is a triplet code, hence each amino acid is encoded by three bases.

18. S. Gillam and M. Smith, "Site-specific mutagenesis using oligodeoxyribonucleotide primers: II In viro selection of mutant DNA," *Gene* 8 (1979): 99–106.

19. M. J. Zoller and M. Smith, "Oligonucleotide-directed mutagenesis using M13-derived vectors: an efficient and general procedure for the production of point mutations in any fragment of DNA," *Nucleic Acids Research* 10 no. 20 (1982): 6487–6500.

20. M. J. Zoller and M. Smith, "Oligonucleotide-directed mutagenesis: a simple method using two oligonucleotides primers and a ssDNA template," *DNA* 3 no. 6 (1984): 479–488.

21. T.A. Kunkel, "Rapid and efficient site-specific mutagenesis without phenotypic selection," *Proceedings of the National Academy of Sciences* 82 (1985): 488–492.

22. G. Winter, A.R. Fersht, A.J. Wilkinson, M. Zoller, and M. Smith, "Redesigning enzyme structure by site-directed mutagenesis: tyrosyl tRNA synthetase and ATP binding," *Nature* 299, no. 5885 (1982): 756–758.

23. G. Weinmaster, M.J. Zoller, M. Smith, E. Hinze, T. Pawson, "Mutagenesis of Fujinami sarcoma virus: evidence that tyrosine phosphorylation of $P130^{gag\text{-}fps}$ modulates its biological activity," *Cell* 37 (1984): 559–568. A kinase is able to add a phosphate residue to a substrate; a tyrosine kinase phosphorylates tyrosine residues in proteins. Many protein kinases are also able to carry out autophosphorylation, i.e. they can phosphorylate themselves. The Tyr in $p130^{gag\text{-}fps}$ would later be shown to be phosphorylated. When the tyrosine is mutated to phenylalanine, the enzyme cannot be phosphorylated at this position.

24. S.L. McKnight, R.C. Kingsbury, A. Spence, M. and Smith, "The distal transcription signals of the herpesvirus tk gene share a common hexanucleotide control sequence," *Cell* 37 (1984): 253–262.; M.R. Briggs, J.T. Kadonaga, S.P. Bell, and R. Tjian, "Purification and biochemical characterization of the promoter-specific transcription factor, Sp1," *Science* 234 (1986): 247–52.

25. A.G. Mauk, "Electron Transfer in Genetically Engineered Proteins. The Cytochrome c Paradigm," *Structure and Bonding* 75 (1991): 131–157.

26. M.H. Caruthers, "Gene synthesis machines: DNA chemistry and its uses," *Science* 230 (1985): 281–285.

INDEX